New Directions in Philosophy and Cognitive Science

Series Editor

Michelle Maiese
Admin 451, Emmanuel College
Boston, MA, USA

This series brings together work that takes cognitive science in new directions. For many years, philosophical contributions to the field of cognitive science came primarily from theorists with commitments to physical reductionism, neurocentrism, and a representationalist model of the mind. However, over the last two decades, a rich literature that challenges these traditional views has emerged. According to so-called '4E' approaches, the mind is embodied, embedded, enactive, and extended. Cognition, emotion, and consciousness are not best understood as comprised of brainbound representational mechanisms, but rather as dynamic, embodied, action-oriented processes that sometimes extend beyond the human body. Such work often draws from phenomenology and dynamic systems theory to rethink the nature of cognition, characterizing it in terms of the embodied activity of an affectively attuned organism embedded in its social world. In recent years, theorists have begun to utilize 4E approaches to investigate questions in philosophy of psychiatry, moral psychology, ethics, and political philosophy. To foster this growing interest in rethinking traditional philosophical notions of cognition using phenomenology, dynamic systems theory, and 4E approaches, we dedicate this series to "New Directions in Philosophy and Cognitive Science."

If you are interested in the series or wish to submit a proposal, please contact Amy Invernizzi, amy.invernizzi@palgrave-usa.com.

More information about this series at
http://www.palgrave.com/gp/series/14744

Konrad Werner

The Embodied Philosopher

Living in Pursuit of Boundary Questions

palgrave
macmillan

Konrad Werner
Faculty of Philosophy
University of Warsaw
Warszawa, Poland

New Directions in Philosophy and Cognitive Science
ISBN 978-3-030-79963-2 ISBN 978-3-030-79964-9 (eBook)
https://doi.org/10.1007/978-3-030-79964-9

Cover illustration: agefotostock / Alamy Stock Photo

This Palgrave Macmillan imprint is published by the registered company Springer Nature Switzerland AG.
The registered company address is: Gewerbestrasse 11, 6330 Cham, Switzerland

Contents

1

Introduction: Philosophizing as a Peculiar Pursuit

1.1 What Is It Like to Philosophize? Toward a Bottom-Up Meta-philosophy

Competence in philosophy can be regarded as a desirable intellectual virtue. Philosophers are skilled at asking questions that usually do not come to mind and clarifying issues that may otherwise be blurred. For example, Michael Dummett notes that while the optician "provides us with spectacles that bring all that we see into sharper focus," the philosopher "aims to perform a similar service in respect of our thinking about reality" (Dummett, 1991, p. 2). As for the "continental" side, Maurice Merleau-Ponty believes that phenomenology (therefore, arguably, philosophy as a whole, too) "slackens the intentional threads which attach us to the world and thus brings them to our notice" (Merleau-Ponty, 1962, p. xiii).

In both cases, the purpose is to elicit the fundamental, yet unnoticed, factors shaping our worldviews. This peculiar skill, the *capacity to problematize*, first in general and then specifically in philosophy, is the subject matter of this book. I shall explore its nature and functioning. This will lead me to a certain conception of why one group of animals, mostly

K. Werner, *The Embodied Philosopher*, New Directions in Philosophy and Cognitive Science, https://doi.org/10.1007/978-3-030-79964-9_1

engaged in dealing with the needs they share with other living creatures, thus in maintaining their precarious autonomy, is *also* engaged in the pursuit of problematizing, including philosophical reflection. I mean humans—one might add "of course" ("Humans, of course!"), but actually there is no reason to do that. In principle, this could have been any lineage.

Now, crucially, this appeal to evolution, this talk of "living creatures," is by no means mere decorum, a fancy way of getting the reader interested. I mean it technically. The aim of this book is to set forth an account of what the pursuit of problematizing in general, and the pursuit of philosophizing in particular is, qua undertaken by a living system coupled with its surroundings (this task is fulfilled in Chap. 5, following various preparatory investigations). Living creatures do various things, not only to survive, but also to *make sense* of the world they live in, as Francisco Varela puts it, and some of them do philosophize. Why? How do they manage it? How is this specific and quite peculiar endeavor related to the way in which the world is experienced by them? What is it like to philosophize? These are our concerns.

The concerns are meta-philosophical, yet in this book they are approached from a perspective that might be called bottom-up, which means that we shall try to reconstruct first how philosophizing emerges from and finds itself within the actual experiences of conscious, self-reflective subjects; and second, how the actual processes of maintaining one's precarious autonomy and one's environmental interactions have given rise to the peculiar intellectual endeavor called philosophy.

1.2 Troubles with Philosophy: The Strategy of Escaping Forward

All that said, philosophizing can also be a risky endeavor due to the highly peculiar nature of philosophical questions and claims. Alongside the appreciation philosophy has been shown, some ingredients of philosophy have also been seen as nonsense. We shall examine these claims, made by Bertrand Russell, Ludwig Wittgenstein, and Rudolf Carnap, among others. The point is not to dismiss them, but to learn from them.

In addition to the arguments put forward by these three early analytic thinkers, mostly referring to logic and the somewhat dubious logical status of philosophical claims, more recently a number of empirical challenges have also emerged thanks to researchers working in the experimental philosophy paradigm. Arguably, the rise of experimental philosophy (hereafter referred as Xphi) along with newly emerging strategies of defending a more traditional way of philosophizing rounded up under the umbrella term "expertise defense" creates a new context in which we should search for a more thorough understanding of philosophical competence.

My working understanding of Xphi is that it aims at eliciting the beliefs of ordinary folks as distinguished from those held by philosophers, but somehow referring to philosophically relevant topics, such as the nature of knowledge or intentional action. For example, Roberts et al. (2017) believe that "if intuitions are relevant to philosophical theories (…), there is reason to be concerned with what non-philosophers' intuitions are" (Roberts et al., 2017, p. 3270).

To get a sense of how Xphi works, imagine that you are presented with a scenario inspired by Edmund Gettier's famous (1963) paper, in which an "everyman" named John believes, for example, that the capital of the United States is located along the Potomac River and one day even voices this belief to a friend, but is not aware of the fact that the night before, the capital was moved from Washington D.C. to Rosslyn, Virginia, just across the river. John's belief is true and he is justified in holding this belief, as he has many times visited the riverside (which is just one of various possible justifications). But can we really say that John *knows* that the capital of the United States is located along the Potomac River in this particular situation, taking into account that while his statement actually is true, he nevertheless has a wrong city in mind? Well, the best thing we can do is to ask people and check out what *their intuitions* about this case are, a proponent of Xphi would say. The idea, indeed, seems reasonable.

There is a large body of literature on intuitions and how a philosopher can rely on them (see, e.g., Weinberg, 2007; Nichols & Ulatowski, 2007; Swain et al., 2008; Kuntz & Kuntz, 2011; Nado, 2014). My topic, however, is different: here the question is not whether or not philosophers can rely on *their* intuitions (if they need to at all; see Cappelen, 2012), but rather what makes a specific cognitive or intellectual faculty, be it intuition or anything else, the driving force behind philosophy. Given that there are philosophical intuitions, beliefs, problems, the question is this: what makes them *philosophical*? So, in my reading Xphi does not only challenge the way in which philosophers use intuitions; it challenges our understanding of what philosophy is (or should be).

Although approaching the pursuit of philosophizing from the naturalistic perspective (thus taking it as something that some animals do as part of their dealing with their environment) may seem somewhat progressive, if I can use this word, the goal of the book is in fact—to stay within the same "political" framework—conservative. For the empirical challenge posed by Xphi seems to me the most pressing issue to be tackled if the intellectual virtue of philosophizing is to be *preserved;* if philosophizing is to stay what it has been for millennia.

It just happens to be the case that upholding this conservative stance these days requires a strategy that is very aptly articulated in the German phrase "die Flucht nach vorn antreten" sometimes translated into English as "taking the bull by the horns," although the latter fails to capture the dynamic quality of the former. Literally the former means *escaping, yet not by taking a backward route, but by taking a forward route*, and thereby confronting the challenge full on; thus—*escaping forward* (which is also conveyed by the Polish idiom "ucieczka do przodu"). In our context, escaping forward means avoiding the adverse consequences of some of the newer ideas while maintaining more traditional views (thus *escaping*), but doing so by engaging in the fabric of these newer ideas, in the fabric of modern cognitive science more generally, and therefore adopting some of them (hence *forward*) for the sake of preserving what we recognize to be most precious about the traditional approach (therefore—*escaping forward*).

1.3 Challenging the Institutional Approach to Philosophy

The intellectual virtue attributed to the pursuit of philosophizing, as well as the nonsense seen to also qualify some of it, has become, in a sense, institutionalized as part of the modern division of labor in academia. This is not bad in itself, of course, but it is one thing to impose an institutional framework on people's activities for practical reasons—society needs certified specialists, including philosophers—and another to *define* these activities solely in terms of their contemporary, in principle changeable, institutional framework.

Relying too much on the current institutional regime may lead one to say that all we have to do in order to distinguish philosophers from non-philosophers is make sure that no philosophy student, graduate, not to mention professor participates in the group we want to target (or that we have professionals only if we want to target them specifically). This claim articulates what I provisionally call the *institutional approach to philosophical competence*. As in the institutional account of the work of art (the classic posit by George Dickie, 1974 based, at least to some extent on Arthur Danto, 1964), here ascribing philosophical competence to a person *s* is not justified in light of any internal features of *s*, but by the position of *s* within certain institutions. Hence, an entity counts as a work of art as long as it is cast in this role or appreciated by an authorized person in accordance with the division of work in a given society (according to Dickie, 1974; Danto's, 1964 view is more complex in this respect), and along the same lines, the institutional approach to philosophical competence recognizes *s* as philosophically competent if and only if *s* has gone through institutionalized philosophy curricula, examinations, and so on, and was then certified by people authorized to issue such certificates.

It is reasonable to assume that taking philosophy as an intellectual virtue has been the driving force behind the mentioned institutionalization. Otherwise, if philosophizing is inherently nonsensical, why should anyone care about its proper placement in academia? Yet, on the other hand, if philosophers are unable to articulate plausible claims and prove theses about reality, as the criticism invoked above aims to show, they can only express their opinions or *beliefs*. Doing philosophy consists of careful analyses of the philosopher's own personal beliefs based on specific intuitions. Now, if we take these two ideas together, we get the institutional approach in its full-fledged guise: philosophers are institutionally certified as specialists in articulating and analyzing beliefs regarding matters that cannot be assessed scientifically. They are *certified believers*, so to speak, as peculiar as it may sound.

The two components of the approach flagged above are therefore as follows: first, the institutional approach which tacitly and somewhat unconsciously follows the pattern championed elsewhere by Danto; second, what may provisionally be dubbed the *doxastic* approach to philosophical competence, making the case that philosophy consists of *beliefs* pertaining to certain abstract matters such as knowledge, truth, or justice. Both of them shall be abandoned, yet in due course.

Chapter 2 provisionally assumes the doxastic approach and seeks to demonstrate its insufficiencies, particularly the fact that it neglects questions that the said beliefs are answers to. Drawing on this recognition, Chap. 3 focuses on questions first, and then on the capacity to ask questions—the capacity to problematize this or that part or aspect of a questioner's surroundings. The fixation on beliefs is not so much criticized—after all, what's wrong with beliefs?—but rather abandoned for the sake of a deeper insight into the origins and mechanisms of problematization as such (Chaps. 3 and 5) and philosophical problematization in particular (Chaps. 4 and 5). As for the institutional component, it is in the air, so to speak, in Chaps. 2, 3 and 4, since we need to have some provisional, default way of referring to philosophers as distinct from people of other occupations, yet it also fades over time so that in Chap. 5 it becomes clear that while referring to the division of labor may serve as a useful approximation or operationalization for empirical study or even armchair speculation, it tells us nothing insightful about who philosophers are and what they do.

1.4 Chapter to-Go

Aside from the Introduction and Epilogue, the book is composed of four chapters which can be read in separation. Each of these chapters to some degree belongs to a different tradition, a different theoretical context, and makes use of (to a large extent) different literature references. The reader may also agree with the conclusions of one chapter but reject the conclusions of the others. The order of the chapters is determined by the logic of my argumentation, the subsequent steps of the investigation as I designed it, but in my subjective assessment it also reflects a certain hierarchy of acceptability. So Chap. 2 features certain proposals that could appeal to a wider group, especially to analytic philosophers, who may not necessarily be eager to embrace the intuition talk of Chap. 3. Moreover, Chap. 4 draws on a very obscure, rather old and abandoned project called the *philosophy of radical beginning*, following Husserl and Carnap (actually also Plato, Descartes, and many others), whose appeal may even be smaller. Finally, Chap. 5 brings up the conceptual framework being developed in most contemporary philosophy of biology and philosophy of cognitive science—referred to as embodied cognition, enactivism, among many other labels. Therefore, accepting the context in which I lay out my final conclusions is in fact wholly independent of what the reader's reaction to Chaps. 2, 3, or 4 might have been.

This way of structuring the book is somewhat odd, perhaps, for it may give the impression of a kind of chapter-to-go system in which, as with coffee to-go, you don't have to go inside, thus really engage in anything. But that is not accidental. The book was designed this way. It enables the reader to take a posit and develop it further in a direction of their choosing without necessarily buying the whole story. I myself am not sure how it will play out, but I also believe that it is worth giving it a chance especially in a time when different philosophical parties rarely refer to one another. For example, the two most consequential reformist paradigms in contemporary philosophy (in my assessment)—embodied cognition and experimental philosophy—maintain no connection whatsoever.

In sum, the book elaborates on the following themes:

- *Comparability of beliefs*. In Chap. 2, I argue that in principle beliefs are answers to specific questions. Sometimes this is apparent, when, for example, someone says something like "I believe that 1505 was the year Poland adopted the *Nihil novi* constitution, not the 3rd of May Constitution." However, the latter indicates not a particular question, but a range of possible questions including "Was it the *Nihil novi* constitution or the 3rd of May Constitution that was signed in 1505 by the Polish parliament?" or "There was a pivotal law written in Poland in 1505, was it the 3rd of May Constitution?" As we can see, even beliefs that unveil some of the content of the queries they answer still remain somewhat undetermined. And what of beliefs that don't even do as much? I argue in Chap. 2, based on various sources, that beliefs should be taken together with the questions they reply to; the latter being represented by sets of alternative replies called contrastive alternatives. Two beliefs are comparable if their contrastive alternatives have common elements. Comparability is a matter of degree. As for the philosophical beliefs expressed by a person acting in the capacity of a philosopher (qua philosopher—expressions like these will also be targeted in Chap. 2) and beliefs that may seem related to philosophy on the surface but are articulated by another person acting in a different capacity (who may be the same person in some cases), they can be juxtaposed only if their comparability has been proven first.

- *Erotetic intuition*. Having established that beliefs should be considered together with the questions they answer, more attention shall be given to questions themselves and how they come about in the first place. In Chap. 3, I examine various theories concerning questions. It turns out that both questions and answers are to some degree semantically undetermined when approached separately, and should therefore be considered together as parts of larger structures composed of *desiderata* (requests for information) and what I propose to refer to as *saturata* (information that answers the request). I then argue in Chap. 3 that it is impossible to account for all desideratum-saturata pairs in terms of instructions and logical operations on propositions. If the whole structure of knowledge built on answers to specific requests for information is supposed to refer to more than the realm of mental representations, we need to postulate a direct erotetic link with our surroundings. In this context I put forward the provisional idea of erotetic intuition: the intuitive capacity to problematize, to invite certain things, metaphorically speaking, to present themselves qua problems. Erotetic intuition, however, is a black box—at this level we have no vocabulary to provide a model of it and therefore intuition talk is supposed to cover our ignorance while at the same time enabling us to at least articulate some introductory insights and move on in the hope of finding a more apt conceptual framework along the way.
- *Competence in philosophy*. Once we have a provisional way of talking about the capacity to problematize—the erotetic intuition talk—we can ask what is special about the acts of problematization pursued by

philosophers. I try to answer this question in Chap. 4, to the extent to which it is possible at all given such limited space. Yet, crucially I don't start with a ready-made list of utterances that are usually assumed to express a philosophical question of belief. Instead, I ask what actually distinguishes a philosophical act of questioning or problematizing from other such acts, independently of how this act is subsequently articulated and of whether the act is pursued by a professional philosopher. I argue in Chap. 4 that what makes philosophical problematizations special is what I call the recognition of one's epistemic position. From the first-person perspective, this means that I *become aware* that the whole manifest image of the world, thus the world as I know it, as I perceive it, is a certain presentation constrained by my cognitive capacities and location. In other words, I discover the seemingly trivial "fact" that not only am I *in* the world, as a rock or a glass of water are, but also the world is *a certain way for* me—something that cannot be said about the rock or water. Partial recognitions of this kind are likely elements of everyday life and they are certainly familiar to cognitive psychologists and other researchers studying mechanisms of perception, categorization, context dependence, and so on. What is special about philosophy, however, is that it strives to take this recognition to the extreme and draw conclusions from the awareness that basically *all* that we encounter and do, including the most familiar things on the one hand and the most sophisticated scientific models on the other, is in one way or another adapted to our cognitive fit. If that is so, there emerges the boundary question about the status of this overall presentation *qua presentation*: what is its connection to a hypothetical reality in itself and others. These are not, and cannot be, scientific questions, but we must determine whether they are sensible in the first place. Chapter 4 provides an account of their role, to be developed further in Chap. 5.

- *The embodied pursuit of problematization.* Finally, given that we cannot account for our general erotetic capacity to problematize, and to problematize philosophically in particular, using standard meta-philosophical methods and standard methods of analyzing questions, I turn, in Chap. 5, in a somewhat less familiar direction. By "less familiar" I mean that it is little familiar to those participating in the debate on the status of philosophy, the relationship between armchair and experimental philosophy, and so on. Outside of these circles, the conceptual framework I lay out is currently becoming more and more popular and widely recognized. I am referring to the cluster of theories rounded up by the umbrella term *embodied cognition* or enactivism. In order to avoid sectarian issues as to whether my posit fits this newly emerging paradigm in the philosophy of cognitive science, and also because I adopt a number of ideas from the philosophy of biology (which maintains loose links with embodied cognition), I have decided to use the label "biological externalism." I build my theoretical construction on several posits taken from the above-mentioned areas, most importantly: the concept of autonomy and the autonomous system; the life–mind continuity thesis; niche construction theory; the concept of the cognitive niche; finally—and crucially—the sensorimotor invariants model. Chapter 5 provides a simple functional model of cognitive niche construction and unpacks the pursuit of problematization in this context. Problematization is taken as a strictly biological phenomenon: the capacity to extract information from relatively stable and remotely predictable parts of the environment that initially—thus prior to problematization—do not convey any information. I argue that while the cognizer, thought of as a living creature, has several strategies of dealing with uncertainty in its environment—which is a widely held view in studies of cognition—the cognizer also needs a cognitive capacity designed to deal with *too much certainty*. This is the role of problematization. Within this context I also unpack philosophical problematization which, as I see it, is the pursuit of problematization in *pure form*, and—in a sense—a natural consequence (a kind of completion) of the evolution of our problematizing skills.

* * *

Acknowledgments This book owes so much to so many people I have met along my way that it would be a daunting task to produce a comprehensive list of names. First of all, I would like to thank my family—my wife and daughter, and my mom—for their patience, presence, and understanding. I am also grateful to my friends who are always there, open to my concerns. Thank you.

Philosophy is something that you really have to learn from other people. Here the first person I need to thank is Jerzy Perzanowski, whom I knew just for two or three years before he passed away in 2009, but it was enough to reset my entire thinking about philosophy. Perzanowski is one of the greatest thinkers of our times in the area of formal ontology and applications of logic to philosophy in the first place, but what I remember most is his incredibly broad approach, capable of taking in the whole intellectual tradition in a single synthetizing act of the intellect.

I would like to thank my colleagues, particularly but not exclusively those from the University of Warsaw, where I teach now and at Jagiellonian University, where I received my original training in philosophy. I am especially grateful to Józef Bremer, Katarzyna Paprzycka-Hausman, Mieszko Tałasiewicz, and Janusz Sytnik-Czetwertyński, who helped me at various stages of my career.

All the mixed and mingled topics touched upon in this book reflect all of my explorations over the last few years in various areas of philosophy, sometimes clearly related to one another and sometimes maintaining a more intricate connection. To list just two that have been most consequential, when it comes to ontology, my gratitude goes especially to Achille Varzi. As for my venture into the area of philosophy of biology, I am especially indebted to Colin Allen.

This book would not be possible without the help of Dominika Gajewska, whose hard work on correcting my English over the course of many years must have been a difficult task, to say the least.

Last but certainly not least, I want to thank the anonymous reviewers of this book and the editors from Palgrave Macmillan for their hard work and encouragement.

The preparation and writing of this book was generously supported by two grants from the Polish National Science Center (Narodowe Centrum Nauki): grant no 2016/20/S/HS1/00046. (2016–2019) and grant no 2019/33/B/HS1/01764 (2020–2023).

It should also be noted that parts of Chap. 4 have already been published in Werner (2020).

References

Cappelen, H. (2012). *Philosophy Without Intuitions*. Oxford University Press.

Danto, A. (1964). The Artworld. *Journal of Philosophy, LXI*, 571–584.

Dickie, G. (1974). *Art and the Aesthetic: An Institutional Analysis*. Cornell University Press.

Dummett, M. (1991). *The Logical Basis of Metaphysics*. Harvard University Press.
Gettier, E. (1963). Is Justified True Belief Knowledge? *Analysis, 23*, 121–123.
Kuntz, J. R., & Kuntz, J. R. C. (2011). Surveying Philosophers About Philosophical Intuition. *Review of Philosophy and Psychology, 2*, 643–665.
Merleau-Ponty, M. (1945/1962). *Phénoménologie de la perception.* Available in English: *Phenomenology of Perception* (C. Smith, Trans.). Routledge & Kegan Paul.
Nado, J. (2014). Philosophical expertise. *Philosophy Compass, 9*, 631–641.
Nichols, S., & Ulatowski, J. (2007). Intuitions and Individual Differences: The Knobe Effect Revisited. *Mind & Language, 22*(4), 346–365.
Roberts, P., Andow, J., & Schmitdtke, K. A. (2017). Lay Intuitions about Epistemic Normativity. *Synthese, 195*, 3267–3287. https://doi.org/10.1007/s11229-017-1371-6
Swain, S., Alexander, J., & Weinberg, J. (2008). The Instability of Philosophical Intuitions: Running Hot and Cold on Truetemp. *Philosophy and Phenomenological Research, 76*(1), 138–155.
Weinberg, J. (2007). How to Challenge Intuitions Empirically Without Risking Skepticism. *Midwest Studies in Philosophy, 31*, 318–343.
Werner, K. (2020). Philosophical Intuition Is the Capacity to Recognize one's Epistemic Position. An Old-Fashion Approach Based on Russell, Carnap, Wittgenstein, and Husserl. *Philosophia, 48*, 1725–1751.

2

Questioning the Comparability of (Philosophical) Beliefs

2.1 The Puzzle of Comparability

Traditional philosophical analyses based on thought experiments are a rich resource for questionnaires. It is tempting to find out what people really think about scenarios like those proposed by Gettier (1963), who challenged the philosopher's understanding of propositional knowledge, or Putnam's (1981) brain in the vat story. Experimental philosophy (Xphi) seeks to satisfy this temptation (see Woolfolk, 2013; Andow, 2016; Baz, 2017 when it comes to Xphi's methodology). It is fair to say that in most cases Xphi tries to elucidate the intuitive beliefs of ordinary folks (taking into account social, cultural, and other variations) as distinguished from those of philosophers, yet referring to philosophically relevant concepts or issues, for example, knowledge (Beebe, 2012; Buckwalter, 2010, 2012; Starmans & Friedman, 2012; Buckwalter & Schaffer, 2015; Machery et al., 2015; Roberts et al., 2017), skepticism (Nichols et al., 2003), or intentional actions (Knobe, 2003).

For the sake of the argument, I will provisionally assume that folks may have intuitive beliefs that somehow pertain to philosophically

K. Werner, *The Embodied Philosopher*, New Directions in Philosophy and Cognitive Science, https://doi.org/10.1007/978-3-030-79964-9_2

relevant issues. The chapter therefore speaks of the intuitive, philosophically relevant beliefs of laypeople—in short *folk beliefs* or *ordinary beliefs*.

It is by no means clear—and this issue is the subject proper of this chapter—whether folk beliefs are in fact *comparable* to the beliefs of philosophers. I do not mean the question of whether the latter are more elaborate or thought through, nor whether the former are interesting to a professional philosopher. I mean the puzzle of whether it is *correct* to *compare* them.

What the comparison at stake, as well as its legitimacy, amounts to is going to be unpacked in more technical terms later on in this chapter. For now, let me illustrate the issue by an example from a non-philosophical area. Suppose I say at a public meeting that I prefer Wagner to disco but the audience does not hear the "to disco" part because the microphone gets broken. Subsequently, someone from the audience lashes out against me, expressing their disdain toward Wagner's pompous style and in turn praising Bach's or Chopin's taste and self-restraint. It seems that our beliefs are in disagreement, while in fact they invoke completely different frames of reference or different *contrasts*, as Jonathan Schaffer (2007) would say. We therefore have *Wagner vs. disco* on the one hand, and *Wagner vs. Bach* on the other. This renders our beliefs incomparable due to accidental causes (the broken microphone) despite the fact that they may otherwise seem comparable. Within a *proper* frame of reference, I would be happy to affirm the superiority of Bach, and my "opponent" could also concede that, after all, despite Wagner's pompous style, it still makes more sense to listen to Wagner than to disco.

Someone might say that what the above scenario illustrates is simply a conversational misunderstanding, but, in my view, it goes much deeper than that, although admittedly the example is quite imperfect. I argue that the dissonance described above pertains to the *structure* of beliefs which should not be taken as isolated utterances, but as parts of larger compounds such as "I believe that it is good to listen to Wagner *qua an answer* to the question of whether people should listen to Wagner or disco" and "the man from the audience believes that it is good to listen to Bach *qua an answer* to the question of whether people should listen to Wagner or Bach." The two beliefs are not incongruous, but rather *incomparable* due to the fact that they stand as constituents of two entirely different alternatives.

I shall base my argumentation on the following two theoretical posits drawn from two entirely different contexts: firstly, on Kit Fine's (1982) theory of embodiment which makes deliberative use of "qua," a traditional logical operator responsible for introducing or unveiling contexts, contrasts, or aspects; and secondly, on the notion of contrastive knowledge developed by Schaffer (2007), who makes the case that our knowledge is populated by propositions (here *beliefs*) relative to fixed questions, rather than isolated propositions on their own.

Therefore, the lesson we need to take away from these introductory considerations is that when seeking to elucidate the status of philosophical competence, we need to begin by refocusing our attention—from beliefs *per se* to the questions they answer. But are there any specifically philosophical questions? From a different angle: are any philosophical questions possible? The latter is the core puzzle which we shall arrive at in Chaps. 4 and 5.

2.2 Comparability in Practice: Some Initial Concerns

The question of comparability should be distinguished from the question of whether laypeople are competent enough to take a position on philosophical issues, as well as whether philosophers are skilled enough to avoid the usual cognitive biases such as the ordering effect (Schwitzgebel & Cushman, 2012) and framing effects (Tobia et al., 2013).

The issue of comparability should also be distinguished from the question of whether or not folks ascribe the same *meanings* to the terms used by philosophers, such as "knowledge" and "intentionality" (I shall address this topic briefly).

One may assume either that the two groups, that is, ordinary people and philosophers, have comparable beliefs or that they have incomparable beliefs. I shall refer to the first claim as the comparability thesis, and to the second as the incomparability thesis. The comparability thesis is virtually never formulated explicitly as a posit that would necessitate separate debate. Rather, it is *assumed* in experimental settings, hypotheses as well as discussions of Xphi. For example, Kuntz and Kuntz state that "intuitions of the folk are inconsistent with the published intuition-data of professional philosophers" (Kuntz & Kuntz, 2011, p. 644). However, it is not clear what the alleged inconsistency means. Recall the claims about Wagner, Bach, and disco invoked earlier—they seem inconsistent, but in fact they are not: they are incomparable, which means that they *cannot* be inconsistent. So, by affirming inconsistency, Kuntz and Kuntz (2011) implicitly affirm comparability. In the same spirit, Starmans and Friedman (2012) point out that while we still know very little about the epistemic attitudes of the folk, philosophers are, so far, the best-studied species when it comes to people's attitudes to knowledge. They write that "we do know quite a bit about how one group of people think about knowledge" (Starmans & Friedman, 2012, p. 273), by which they mean philosophers, who "have inquired into the nature of knowledge for millennia" (ibid.). The assumption seems to be that what philosophers "think about knowledge" can be confronted with what others "think about knowledge," and thus that this juxtaposition is itself correct. But what does it mean to *think* about knowledge? Does it mean the same thing for both groups? When it comes to philosophers, we know what this or that philosopher *claims* with respect to knowledge in abstracto, thus as part of a theory. However, aside from the specific issue of knowledge, thus more broadly, is there any reason to believe that this kind of "thinking," that is, these theoretical claims, have anything to do with what non-philosophers think, state, and do?

Let us deal with one example in greater detail, namely the experiment ran by Nichols et al. (2003) which became a classic in the field of Xphi. The participants were presented with the story, based roughly on Putnam (1981), about George and Omar, who are roommates and enjoy having philosophical discussions. One night Omar argues:

> At some point in time, by, like, the year 2300, the medical and computer sciences will be able to simulate the real world very convincingly (…) so that the brain has experiences exactly as if it were a real person walking around in a real world, talking to other people, and so on. And so the brain

> would believe it was a real person walking around in a real world, etc., except that it would be wrong (…). And here's the thing: how could you ever tell that it isn't really the year 2300 now, and that you're not really a virtual-reality brain? If you were a virtual-reality brain, after all, everything would look and feel exactly the same to you as it does now! (Nichols et al., 2003/2011, p. 240)

After reading the vignette, the participants were asked to give their assessment of George's epistemic position and to decide whether he REALLY KNOWS or ONLY BELIEVES that he is a real person. It turned out that the "low philosophy" group, that is, participants lacking philosophical competence (measured as the function of how many philosophy courses they had taken), were more likely to choose REALLY KNOWS (about 50 percent picked this option) than the group dubbed "high philosophy" (only about 20 percent, while about 80 percent picked ONLY BELIEVES).

Clearly, the answers provided by the "low philosophers" were targeted along the very same lines as those given by the "high philosophers," meaning that the two groups of answers are such that, according to Nichols et al. (2003), *it makes sense* to confront them and say something like "look, the 'high philosophers' claim this, whereas the 'low philosophers' claim that."

Someone might argue that the confrontation in question makes sense only if the two groups ascribe the same meanings to "REALLY KNOWS" and "ONLY BELIEVES," which is doubtful, taking into account precisely the differences in rates of philosophical study: what philosophers mean by "REALLY KNOWS" and "ONLY BELIEVES" is loaded with content that is not familiar to folk, and this fact amounts not only to a difference in the degree of both groups' familiarity with the said phrases, but to the assignment of two genuinely different meanings to the phrases in question.

Yet, as noted earlier, questioning comparability differs from the argument pertaining to alleged differences in meaning. Note that two (or more) philosophers may also really differ when it comes to the meaning they ascribe to "REALLY KNOWS" and "ONLY BELIEVES," if, for example, they subscribe to different philosophical traditions (e.g., one of them considers themselves a follower of Russell, while the other adheres to the legacy of Hegel and Bradley). But does it mean that they ascribe two (or more) different meanings to the said phrases? If they did, it would make philosophical debates virtually impossible.

Consider the following elaboration of the original case. A simplified map of beliefs or attitudes can look as follows: Strong account of knowledge (*S*) answers "ONLY BELIEVES" to the Putnamian story of George and Omar, which means that there are certain demanding standards that have to be satisfied by any belief if it is to be deemed knowledge, and our ordinary beliefs frequently do not stand up to them (this is the Cartesian tradition, broadly construed). Weak account of knowledge (*W*) denies *S* and says "REALLY KNOWS," meaning, perhaps, that knowledge is a more practical concern, and therefore, a huge group of everyday beliefs count as knowledge. People who are philosophically competent are therefore, according to Nichols et al. (2003), more likely to choose *S* than *W*, while folks prefer *W*.

Now, imagine a person, let us call her Jane, who starts from *W*. She is naive, meaning that she lacks philosophical competence. Let's then mark her naivety by means of the label "W_n" which is supposed to convey the message that Jane is not committed to any stronger standards of knowledge simply because she has never reflected on the topic. But suppose that as result of studying Descartes and other thinkers she transfers to *S* which may likely be said to be opposed to W_n. Let us also make clear that *S* is *a philosophical* position by marking it "S_p". Now, Jane can take a step forward in her philosophical education and become aware of a variety of problems that arise in connection with such a strong epistemic attitude, and finally, after some lectures she may (yet does not have to, of course) end up being influenced, say, by James, Wittgenstein, Merleau-Ponty or Putnam to the effect that she comes *back* to *W*. But this time she has a sophisticated explanation for this shift. Namely, she concludes, for example, that we must not purport to adopt a God's Eye perspective, as Putnam (1987) called it; that is, we must not cast ourselves in the role of external viewers of worldly goings-on, since such a vantage point is not accessible to us. Therefore, the only thing we can do is sympathize with the characters from the scenario presented by Nichols et al. (2003), and take *their* perspective. Hence—Jane continues—since all the characters populating such stories are fated to have only limited access to information, we must stick to *W* if anyone is supposed to know anything. Otherwise—Jane's argument concludes—"REALLY KNOWS" refers to some extramundane ideal, useless in real-life situations and even in science which is also unable to reach this extramundane perspective (noteworthily, we shall go back to the issue raised by our fictional Jane in Chap. 4). So, Jane ends up holding *W*, but a philosophically refined version of it, so to speak; hence she ends up with W_p, to use the same convention.

What happens between W_n and S_p, and then between S_p and W_p? Finally, what is the difference between W_n and W_p? It *seems* that there must be a difference between the weak epistemic expectations of someone who has never reflected on knowledge (W_n), and the weak epistemic expectations of someone who has gone through the whole history of philosophy and *decides* to *revert to* the everyday naive perspective (e.g., in step with Merleau-Ponty's, 1962 postulate). Yet, W_n and W_p come up in the guise of identical answers when we look through the standard tests used by Nichols et al. (2003).

If the difference between W_n and W_p boils down to a difference of the meanings attributed to "REALLY KNOWS" and "ONLY BELIEVES," then why are we not equally prone to say that S_p and W_p also differ with respect to the meanings of the two phrases? After all, if the former employs the God's Eye perspective and the latter rejects it, there appears to be a major difference between the meanings S_p and W_p attribute to "KNOWS" and "BELIEVES."

In my view, as I shall demonstrate below, drawing on Schaffer's (2007) account of contrastive knowledge, S_p and W_p place both answers "REALLY KNOWS" and "ONLY BELIEVES" in, *ceteris paribus*, the same contrasts, which makes them comparable, and which may not be the case when it comes to W_n. The contrast at stake can be represented as a set of alternative answers to a given question. Hence, in short, S_p and W_p differ as providers of answers to the same set of questions, but they do not differ when it comes to the questions as such. Meanwhile, arguably, W_n—as understood by a non-philosopher—may in fact be answering a different question.

I shall develop the above comparability/incomparability issue in greater detail in the next two sections and in Chap. 3.

2.3 Expertise Defense, Reflection Defense, and the Like

The experiment sketched out above is one of several studies dedicated to making the case that laypeople may have very different intuitive beliefs with regard to philosophically relevant issues; that the beliefs may vary across cultures, socioeconomic status, education, or gender (see Machery, 2017 for a useful summary of the state of the art). Therefore, they argue, it is doubtful whether professional philosophers can rise above all of these individual constraints and develop a better grasp of the issues in question. Roughly, this is the position referred to in the literature as the negative program of Xphi, which argues against attributing any kind of special status to traditional "armchair" elaborations of philosophical intuitions.

Now, a specific denial of the comparability thesis has been formulated in reaction to the negative program. It has come to the fore in the guise of the so-called expertise defense (see Williamson, 2007, 2011; Ludwig, 2007). Different versions of it "contrast the intuitions at play in philosophical discourse with the quick, spontaneous responses provided by non-philosophers to experimental surveys" (Egler & Ross, 2018, p. 3, preprint). However, speaking of "contrast" suggests that the two groups of intuitive beliefs are somehow of the same kind, and thus that the difference between them boils down to the fact that the beliefs of philosophers are based, for example, on better "reflective practices" (Kauppinen, 2007). Here expertise defense falls in line with the comparability thesis laid down in Xphi (for more on the expertise defense see, e.g., Rini, 2015).

However, expertise defense can also be seen in the following way:

> The surveys conducted by experimental philosophers have (…) employed non-philosophers as subjects, but, it is claimed, the intuitions of such persons are irrelevant to philosophical theorizing. (Nado, 2014, p. 632)

Actually, there are some internal differences within the broader tent of expertise defenders, aptly elaborated on by Machery, 2017, but let us work with a somewhat simplified story here. This way or another, the defenders cited by Nado (2014) try to make the case that the answers given by folks are *irrelevant* to the philosophical endeavor proper; in other words, that either the comparison at stake is merely uninteresting (weak reading), yet correct, or that the comparison can never be the case (strong reading). So, while proponents of Xphi, especially practitioners of the negative program, assume that the comparability thesis applies universally, expertise defenders, at least in the more interesting strong version, lean in the exact opposite direction, claiming that comparability is never the case, thus that it should be universally denied.

However, Xphi as well as its critics seems to forgo more detailed scrutiny of what the said comparison as such could be, that is, what conditions would have to be satisfied to make such a comparison possible. Thus, I draw on the diagnosis that both Xphi *and* the expertise defense lack a more scrupulous conceptual analysis of the issue of comparability. I wish to fill this gap at least in part, which shall serve as a stepping stone for further discussion of the issue of questions and the capacity to problematize.

2.4 Different Embodiments of the Subject

As I already mentioned, I shall employ two theoretical posits. The first one is Fine's (1982) theory of embodiment; the second is Schaffer's (2007) theory of contrastive knowledge. I will start with Fine's proposal. Here is its central idea:

> Given any object x and description (property) φ possessed by x, we shall suppose there is a new object x *qua* φ or x *under the description* φ. Thus if x is Socrates and φ the property of being a philosopher, then the new object is Socrates qua philosopher (…).
>
> Given an object x qua φ, we shall call x the *basis* and φ the *gloss*. The resulting object itself will be called a *qua object*, and the operation by which it is formed *glossing*.
>
> One should not identify a qua object with its basis—Socrates qua philosopher with Socrates (…). Rather, the qua object should be regarded as some sort of amalgam of the given object and the property, like the given object but wearing the property on its face. (Fine, 1982, p. 100)

The philosophical challenge that the notion of qua object is originally designed to take on has to do, firstly, with the issue of matter and form, thus with whether an object can be identified with its matter; and, secondly, with the fact that we can attribute different, sometimes contradictory properties or descriptions to seemingly one and the same thing or person (for more on "qua" see Bäck, 1996; Poli, 1999; Werner, 2020). To illustrate the latter claim, take Donald Trump the businessman who can invest in Russia and Donald Trump the president who can't. When we approach the topic in terms of Fine's qua objects we have Trump *qua president of the United States* and Trump *qua businessman*, and while it is perfectly fine for the latter to launch a construction project in Moscow, it is definitely not an option for the former. So, we have two different embodiments of one man and his actions. And these two "persons" can be in a real conflict of interest although they are based in the same living individual.

Several principles govern qua objects according to Fine, determining their identity and the inheritance of properties by qua objects from their bases, but not vice versa.[1] Fine needs the qua object to unpack and solve the ontological puzzle of identity, whereas I am more interested in how different properties/descriptions or capacities can be attributed to what seems to be one and the same thing via a plurality of qua objects it serves as the basis for. The issue I wish to explore, however, does not pertain to cases in which the basis is ascribed opposite or contradictory properties under different aspects, for example, John *qua economist* is rational but John *qua politician* is irrational. Instead, I shall draw attention to the fact that different qua objects can place their common basis in different "spaces," as it were, as already alluded to in the Trump example, to the effect that the ascribed properties/descriptions do not merely "get along" badly (e.g., "rational" and "irrational"), but rather are of different *types*. For example, when I say that "Donald Trump qua *homo sapiens* belongs to the class of mammals" and "Donald Trump qua president is the head of the executive branch of the US government" I place Mr. Trump inside two in principle different frames of reference, or within two entirely different conceptual frameworks: one is the natural-kind classification of species and classes of animals; the other is the logical space established by the American constitution. Now, if someone would hold that Trump *qua president* is a mammal, this would be, I argue, incorrect despite the fact that Trump is president, he is a mammal, and moreover, to my knowledge, all presidents have been mammals so far. Admittedly, someone might argue that the proposition is true, period: if Trump belongs to the group of presidents (true), and the latter group is a subset of a larger group of mammals (true), then "Trump qua president is a mammal" is true. However, it may also be considered (which is the reading I support) neither true nor false, but a nonsensical utterance. I propose that the nonsense is due to the placement of Donald Trump inside two *in principle* different segments of the space of reasons, so to speak.

Now, once the conceptual scaffolding taken from Fine's theory of embodiment has been provisionally set up, we can make our first attempt to characterize comparability in the relevant terms. Namely:

(1) Two beliefs p_1 and p_2 held by the subject s are comparable if there is such φ that s *qua* φ believes that p_1 and s qua φ believes that p_2.

This means that two beliefs are comparable if they can be attributed to a subject taken in the same capacity, thus to a single qua object based on the subject. This basic intuition could be further developed so that we have different subjects and—*ceteris paribus*—identical or analogous glosses, which would make their views comparable.

That said, one way of arguing against the comparability thesis referring to the *philosophers vs. non-philosophers* controversy in terms of qua objects could be to say that there is a fundamental logical difference between attributing beliefs to the subject in her capacity of being a philosopher and attributing beliefs to the subject (even the same subject) in her capacity of being a non-philosopher. Going back to the example rolled out above, we can have:

(2) Jane *qua non-philosopher* holds W (which is referred to as W_n in this case)
(3) Jane *qua philosopher* holds S (which is referred to as S_p)
(4) Jane *qua philosopher* holds W (which is referred to as W_p in this case)

In this account, the difference between W_n and W_p boils down to the difference between the two qua objects standing in the role of the believer in (2) and (4).

Now, I propose that, first, the comparability thesis reformulated in this context is the claim that the very same truth conditions apply to (2) and (4) except for the fact that in (4) Jane must be a philosopher, and in (2) she must be a non-philosopher. In other words, according to the comparability thesis so rendered, the truth conditions *do not change* fundamentally. Suppose that $c_1, c_2, \ldots, c_n$ are the truth conditions of (4) and c_1 = "Jane is a philosopher." The comparability thesis here holds that if we substitute "qua philosopher" with "qua non-philosopher," c_1 shall be replaced with $\sim c_1$ = "Jane is not a philosopher" while the conditions $c_2, \ldots, c_n$ stay exactly the same.

The strong *incomparability* denies the latter point. A suggestion of this kind has been explicitly flagged by Regina Rini:

> It is problematic to compare philosophers and non-philosophers directly on these tasks, because the two groups do not engage with the stimuli in the same way. For non-philosopher subjects, the cases and principles employed are likely to be quite unfamiliar; intuitive judgments and reasoning about them are likely to be experienced as relatively novel cognitive projects. But for philosophers, this is quite unlikely to be true. The case-types and principles employed as stimuli are well-known to philosophers. Philosophers will have memories of previously encountering the stimuli, and how they have responded in the past. So when philosophers react to these stimuli, they are doing something different from non-philosopher subjects, a different sort of cognitive task. (Rini, 2015, p. 44)

Therefore, the strong incomparability thesis would insist that the "qua philosopher" gloss signifies a different attitude toward judgments, beliefs, or "cognitive projects," as Rini (2015) puts it. The "qua philosopher" gloss, therefore, implies that "*x qua philosopher* believes that *p*" has specific truth conditions that are fundamentally different from the truth conditions attributable to "*x qua non-philosopher* believes that *p*."

I will argue in the next section that there is a plausible idea laid down in the expertise defense so unrolled, but unfortunately it is hardly acceptable in its present form. This is because *both* the comparability thesis endorsed by Xphi and the incomparability thesis spelled out above share the same two flaws:

- They take the *philosopher vs. non-philosopher* distinction as mutually exclusive, which may not be the case.
- Both sides of the debate, each one in its own manner, do not sufficiently appreciate the possible differences between grades or levels of scrutiny on the side of non-philosophers: while the expertise defense considers all the beliefs related to philosophically relevant issues held by non-philosophers as not profound enough, Xphi in some sense follows the same pattern and gathers all of these beliefs under the single umbrella term "folk."

2.5 Against the *Philosophers vs. Non-philosophers* Dichotomy

To begin with, even if we assume that it is perfectly clear who a philosopher is, nothing follows from that about who a non-philosopher could be, aside from the conclusion that it must be somebody else: a mathematician, a policeman, an artist, or a miner, just to list some professions. However, note one subtle issue: there is nothing logically wrong with being a philosopher *and* a miner, while it seems, at least at face value, that one can no longer be a non-philosopher once one starts to be a philosopher (unless some additional conditions apply). This dichotomous account raises questions like why are philosophers frequently equally biased in their spontaneous accounts of philosophically relevant situations as non-philosophers, which is taken as a plausible problem and has recently been empirically investigated, for example, by Schwitzgebel and Cushman (2015). In the same spirit, Schwitzgebel (2009) and Schwitzgebel and Rust (2016) found "ethicists neither more nor less likely than other professors to act in accord with their expressed moral attitudes" (Schwitzgebel & Rust, 2016, p. 225). This statement suggests that once one counts as a philosopher/ethicist, one does not count as an ordinary agent any more, which is not at all obvious.

Hence, answering the questionnaire in the capacity of a philosopher implies, according to the reading implicitly employed by many Xphi researchers as well as some critics of Xphi, that one does *not* answer the questionnaire in the capacity of a non-philosopher, and vice versa. Meanwhile, I would hypothesize that if we took a miner who later became a philosopher, and asked him whether he *knows* that it is safe to go underground in such and such conditions, he would *still* answer this question *qua miner*. It is clear that these two glosses, namely "qua miner" and "qua philosopher," place the mentioned individual in two completely different situations or environments. And precisely because they do so, it makes little sense to set them in opposition. They can be at play simultaneously, depending on the context.

Going back to Jane's story, I propose that a better way of unpacking it distinguishes between her capacity as a philosopher, acquired by means of plunging into the works of Plato or Wittgenstein, and her capacity *as a person* or, perhaps even better, *as an agent*, acquired in the course of ontogenesis, upbringing, and social interactions. And here, too, as in the miner case, we have two different locations of Jane which she can maintain simultaneously: *qua agent* she is placed in her life-world, full of perceived things, practical goals, and so on. Meanwhile, *qua philosopher*, she is placed in the logical space of concepts, problems, arguments, and theories developed over the centuries by the group conventionally labeled "philosophers." This distinction also seems natural in light of some experimental results:

> While there have been relatively few studies conducted on professional philosophers, there is some empirical precedence supporting another kind of gap between philosophers' conceptions and their actual practices. (Buckwalter, 2012, p. 519)

Buckwalter is referring to the already mentioned studies by Schwitzgebel and collaborators, yet I would argue that there is no reason to target the gap between what philosophers *think* and what they *do* necessarily as a particular flaw they have (again, it depends on context—it may be seen as a flaw in particular circumstances). Rather, it is an indication of the fact that one and the same person can function—by means of different glosses, thus as the base of many qua objects—in various frames of reference or in many different capacities. Being a philosopher does not mean that one cannot answer a question in a different capacity, not to mention acting in a different capacity.

Crucially, even if a person, for example, Jane in the above story, can only answer a particular question at a given moment in just one capacity, it seems to still be possible for her to "wear" the two glosses, that is, "qua philosopher" and "qua person/agent" on her face interchangeably to the effect that Jane qua philosopher holds S_p (as she did for a moment in the story) and Jane qua agent *at the same time* still holds W_n. In such a case, the gap mentioned by Buckwalter would not boil down to failed implication from philosophical expertise to everyday practice, but—as a gap, indeed—it would mean that there may be no link whatsoever between these two areas.

Therefore, it seems that the gloss "qua non-philosopher" is a somewhat loose umbrella term, and instead of being opposed to this umbrella term, the gloss "qua philosopher" should be seen against the background of a variety of glosses such as "qua agent," "qua person," "qua organism," or "qua miner," "qua mathematician," and so on. Some of these can apply to the subject at the same time as the gloss "qua philosopher."

Now, the expertise defense tries to make the case that philosophers are somehow specially trained in detailed, thorough examination of intuitions referring to such notions as knowledge and intentional action, which is in turn supposed to make them more careful when employing these terms in real-life situations. Williamson (2007) speaks of "careful attention" to subtleties of conceptual analysis, while Kauppinen (2007) appeals to the "reflective practices" of philosophers. Generally, "defenders of expertise suggest that philosophers possess special skill, which enables them to apply concepts to particular cases more carefully and precisely than the folk" (Nado, 2014, p. 632). However, the notion of careful attention as well as the qualifiers "carefully," "precisely," and so on is left rather vague in these contexts. What counts as careful employment of, for example, the notion of knowledge? If we recall the miner again, and his potential knowledge that, for example, the concentration of methane (which is monitored by several devices) at a given moment makes it safe to go underground, I wouldn't dare say that he is *not* sufficiently thorough in his examination of the situation. Analogously, think of a pilot going through a checklist before every take-off and landing. Or think of seventeenth-century Dutch painters and how they studied visual perception, thereby questioning the very line that we, folks (meaning *non-painters*), draw between what things (allegedly) "really" look like and what they seem or appear to look like. All these people should by no means be deemed careless in their assessments of what they know or see. They are all exceptionally careful and precise; philosophers have no monopoly on precision and preciseness in epistemically relevant contexts.

Someone might reply that the latter remark about thoroughness and carefulness is in principle misguided and even unfair because it should be obvious upfront that what the expertise defense means by these categories applies solely to philosophically relevant beliefs and is not meant to deny anybody's intellectual virtue as applied to other fields.

Fair enough, but, firstly, I have presented real-life cases with a significant philosophical component such as knowledge (the miner and pilot) or perception (the painter), just like the cases usually provided to participants in Xphi studies. So, where do we draw the line between the specifically philosophical intellectual skill with respect to knowledge and the skill possessed by other agents who may also have an interest in being extremely careful in assessing what they know? We shall discuss this line in Chaps. 4 and 5. Meanwhile, let us clarify what all this means for the issue of comparability.

2.6 Beliefs and Questions

The condition of comparability given in (1) can still have some appeal, for it does not determine upfront that different embodiments of the same subject fit the dichotomous logic of the *philosopher vs. non-philosopher* distinction. We can still demand that for two beliefs to be comparable, they must be produced and articulated by one or two subjects acting in the same capacity or playing the same role. However, we need to elaborate in more detail on what it means to be acting in a given capacity. It turns out that the very capacity or role in which the subject acts can also be unpacked in terms of qua objects based this time not on the subject, but on her acts themselves.

To begin this part, I would like to bring up the second theoretical posit that my argument is built upon, namely Schaffer's (2007) idea of contrastive knowledge. At first glance it has nothing to do with the issue being debated here, but I shall try to draw a clear connection shortly.

Schaffer starts by characterizing what he calls knowledge-*wh*, which refers to cases such as my knowledge of *who* wrote "The Brothers Karamazov," *whether* Dostoyevsky or Gogol wrote "The Brothers Karamazov," *what* happened on September 1, 1939 in Europe, *when* the Second World War started, *where* I parked my car, *why* the crisis of 2008 happened, and finally also (although it does not fit the *wh* template so well) *how* the brain manages to combine data from two eyes. What is specific in all of these examples is that the *interrogative component*, or, simply speaking, the question that each of these knowledge ascriptions answers, is built into the structure of the ascription. For example, when I say I know when the Second World War started, the question was "When did WWII start?"

Now, the traditional distinction between knowledge-*how* and knowledge-*that*, which we owe to Ryle (1949), appears much more coarse-grained, and it is crucial to Schaffer not to reduce the variety of knowledge-*wh* ascriptions to propositional knowledge-*that*. The reason for the irreducibility of the former is precisely that they have the built-in questions, whereas in knowledge-*that* the question "goes missing," as Schaffer (2007, p. 384) puts it. In other words, when we have the structure K*sp*: "*s* knows that *p*," there is no place for a reference to what question *p* answers. Meanwhile, there could be situations in which the reference helps us figure out what *p* actually says, thus what *s* actually knows.

To demonstrate his point, Schaffer proposes to consider the example of p_1 = "There is a goldfinch in the garden," and therefore Ksp_1 = "*s* knows that there is a goldfinch in the garden." Now, Ksp_1 can be the answer to more than one question. Just consider these:

Q1: Is there a goldfinch in the garden, or a raven?
Q2: Is there a goldfinch in the garden, or a canary? (Schaffer, 2007, p. 388)

Now, Schaffer points out that the two knowledge ascriptions built upon Q1 and Q2 are nonequivalent:

> Knowing whether there is a goldfinch in the garden or a raven is a relatively *easy* task of *bird-identification*. Virtually anyone (with decent vision and minimal expertise) can know whether there is a goldfinch in the garden or a raven. In contrast, knowing whether there is a goldfinch in the garden or a canary is a *harder* task of bird-identification. Perhaps only an expert birder will be able to tell the difference. (…) So one might well know whether there is a goldfinch in the garden or a raven, but fail to know whether there is a goldfinch in the garden or a canary. (ibid.)

The point is, therefore, that the K*sp* scheme works well only in situations when people already know the appropriate interrogative context, thus when the appropriate question is in the air, so to speak. But K*sp* does not make what is in the air explicit. Meanwhile, in order to articulate the question as part of knowledge ascriptions Schaffer proposes his idea of contrastive knowledge-*wh*:

> Here the knowledge relation is relativized to the question: K*spQ* rather than K*sp*. Ksp*Q* may be read as: *s* knows that *p*, as the true answer to *Q*. It is not enough to know that *p*—one must know *p* as the true answer. In other words, it is not enough to know the proposition that just so happens to be the answer—one must know the answer *as such*. (ibid., p. 392)

An alternative way of articulating K*spQ* makes use of the fact that *Q* can also be spelled out in the form of disjunction of alternatives, to the effect that, for example, knowing that Boleslaw Prus wrote "The Doll" as the true answer to the question of whether Boleslaw Prus or Henryk Sienkiewicz wrote "The Doll" is equivalent to knowing that Prus *rather than* Sienkiewicz wrote "The Doll." That is why the knowledge in question is referred to as *contrastive*—each knowledge ascription is contrasted with possible alternatives. Noteworthy, Schaffer's proposal harks back to

Charlers L. Hamblin (1958), one of the founding fathers of erotetics (the logic of questions), who postulates, among other things, that possible answers to a question are an exhaustive set of mutually exclusive possibilities (more in Chap. 3).

Now, my idea is, first, to employ Schaffer's concept to *beliefs*. Thus *p* may be deemed a belief and the structure "*s* knows that *p* as the true answer to *Q*" is naturally transformed into "*s* believes that *p* as an answer to *Q*." The reference to truth is dropped for obvious reasons—knowledge must be true, at least according to the classic definition, but beliefs can obviously be false. Nonetheless, the point is that beliefs, whether true or false, are also bonded closely with relevant questions. Now, I do not claim that all beliefs are answers to questions. Perhaps my belief that rain makes me wet simply stems from experience and I have never asked myself *whether or not* rain makes me wet. However, our attention here is limited to beliefs that do answer questions.

We can therefore try to characterize comparability in terms of contrastive alternatives. Let me begin with an example. Suppose it is November 2020 and John believes that Donald Trump will win the presidential election, while his friend Max believes that Donald Trump has lost the trust of voters in Pennsylvania. Based on the latter, Max believes that Donald Trump will not win the 2020 election. It is clear that the two inconsistent beliefs pertaining *explicitly* to the 2020 election are comparable since they are answers to the same question. They can be unpacked as "John believes that Donald Trump will win the 2020 election qua an answer to the question of whether Trump or someone else is going to win in 2020" and "Max believes that Donald Trump will not win the 2020 election qua an answer to the question of whether Trump or someone else is going to win in 2020." Let us, however, consider the belief pertaining to Trump losing Pennsylvania, expressed by Max. Is the latter comparable to John's belief in Trump's victory? Suppose that it can be rolled out as "Max believes that Donald Trump has lost the trust of voters in Pennsylvania qua an answer to the question of whether Trump has lost the trust of voters in a key state, whose support is necessary for Trump to be reelected." Although at face value these two beliefs are about different things—one is explicitly about the 2020 election and the other is about Trump's polls in Pennsylvania—they can be deemed comparable *and inconsistent* as long as we concede that the following statements are assumed to be true by both John and Max: firstly, "for any candidate, if the candidate loses the trust of voters in any state, she or he cannot win the state;" secondly, "for any candidate, if the candidate loses in California, she or he must win all the key states in the Midwest;" and finally, "Trump will lose in California." Based on that, the belief that Trump will lose in Pennsylvania is a consequence of the belief that Trump has lost the trust of the Pennsylvania voters. All this implies the belief that Trump will *not* win in 2020. Thus the exact negation of John's belief is a consequence of Max's belief pertaining to Trump's failure to deliver Pennsylvania coupled with his earlier failure to deliver California. This makes the beliefs "Trump will win the 2020 election" and "Trump lost the trust of voters in Pennsylvania" comparable despite the fact that they seem to be referring to different issues.

Hence, the situation with John's and Max's beliefs is such that their contrasts *taken together with their consequences* are parts of a broader complex of beliefs, and, as such, these two contrasts have common elements.

Based on this, we can conclude provisionally that:

(5) Two beliefs p_1 and p_2 are comparable if the set of contrast beliefs attributable to p_1, taken together with all its consequences (formally speaking, closed under logical consequence) has a common element with the set of contrast beliefs attributable to p_2, also taken together with all its consequences.

Crucially, comparability so conceived should be considered a matter of degree. One of its extremes is when the two sets of contrasts are identical. In such a case, the two beliefs are perfectly comparable. If I believe that Prus rather than Sienkiewicz wrote "The Doll," and my friend claims that Sienkiewicz rather than Prus wrote "The Doll" (the set of alternatives has just two elements), our claims can be juxtaposed perfectly. The other extreme is when the two sets of contrasts, potentially quite large, have no common elements whatsoever. In the latter case the beliefs are absolutely incomparable.

2.7 Different Embodiments of "Philosophical" Acts of Believing

Shaffer himself takes advantage of the little word "as" in his characterization of knowledge-*wh*, thereby inviting a more elaborate usage of it, thus Fine's theory of embodiment built upon the Latin cousin of "as," that is, "qua."

Fine (1982) makes it clear that the notion of qua object applies not only to material things, but also to acts, and has recently developed a theory of the embodiment of acts (Fine, 2018). The issue that Fine tackles once again has to do with the question of identity. Imagine John who shoots his gun, the bullet reaches his boss' chest and the boss dies a minute later. Was *the act of killing the boss* identical to *the act of shooting the gun*? Someone might argue that they were the same thing, yet John could have no intention to kill, and more importantly, the boss died a minute later, thus how could it be that the two temporally distanced and discrete events constitute one act? Fine proposes to consider certain acts as bases for qua objects, for example the concrete act of John's pulling the trigger taken as a physical event, a bodily movement, can be deemed the basis of the qua object *John's pulling the trigger qua killing the boss.*

Note again, aside from the metaphysical notion of identity which Fine originally focuses on, that different glosses can situate the *base act* within very different frames of reference. While the act of pulling the trigger *qua killing the boss* qualifies the act morally and legally, the act of pulling the trigger *qua ultimate cause of the boss' high-velocity bullet wound* qualifies the act within the framework of medical description without defining its normative status.

Yet, if we are to apply Fine's theory of embodiment to beliefs, we need to make a certain terminological move which is going to prove beneficial later for other reasons, too. Namely, I propose to speak of *acts of believing that p* in addition to the static state of *possessing* a belief *p*.[2] The act of believing is taken as primitive. Beliefs are *products* of acts of believing (following the classic terminology proposed by Kazimierz Twardowski, 1911/1999). As such, acts of believing can be bases of qua objects: we can speak of the act of believing that *p* qua answering the question of whether *p* or *q*, for example.

Putting Fine's and Schaffer's proposal together, we can unpack beliefs-*wh* in terms of qua objects. For example: *s*'s act of believing that Prus wrote "The Doll" qua answering the question of whether Prus or Sienkiewicz wrote "The Doll," differs from *s*'s act of believing that Prus wrote "The Doll" qua answering the question of whether Prus or Joseph Conrad wrote "The Doll." Note that, just as it was the case in Schaffer's example of whether it was a goldfinch or a raven that was in the garden, the first belief (act of believing) referring to Prus poses a relatively harder task than the second if one is not that familiar with nineteenth-century Polish literature. Note in this context that the contrastive approach makes ample room for an account of expertise, as pointed out by Schaffer himself. He stressed that "who has evidence, expertise, and answers is a function of the question." For example, if the question refers to a goldfinch or a raven, "then the person with the evidence, expertise, and answer is the person who can tell a goldfinch from a raven." If the question is easier and involves a raven versus a canary, "then the person with the evidence, expertise, and answer is the person who can tell a goldfinch from a canary" (ibid., p. 394).

So, based on Fine and Schaffer, we obtain something along the following lines:

(6) If s_1 believes that p_1 *qua answering the question of* whether p_1 is the case, or p_2 is the case,…, or p_n is the case, and s_2 believes that q_1 *qua answering the question of* whether q_1 is the case or q_2 is the case,…, or q_n is the case, then the two acts of believing are comparable if (*i*) there are p_i and q_j such that $p_i = q_j$; or (ii) the latter condition is satisfied by some logical consequences of both sets of contrast beliefs.

Note that if $s_1 = s_2$, we can speak of one and the same person having beliefs that are embodied differently, thus are attributable to different contrasts. This proves useful as we come back to the issue of the comparability of philosophical and folk beliefs.

Noteworthily, Schaffer's focus on questions harks back to the idea set forth by Henryk Hiż, who was one of the thinkers who laid the groundwork for logical investigations into the nature of questions (for more on the logic of questions see Wiśniewski 1995, Harrah 2002 and Chap. 3). In the opening statement of his (1962) paper we read:

> Knowledge can be classified according to what questions it answers. (…) That scientific knowledge is systematic may be viewed as a result of systematically arranged answers to systematically arranged questions. (Hiż, 1962, p. 253)

In this context, my main posit is that it is possible (and potentially plausible) to apply the question-based approach to philosophical claims. So, what goes for ordinary beliefs as well as the beliefs proven true (or false) by scientists also goes for beliefs or acts of believing in philosophically relevant cases. In effect, philosophical beliefs must be targeted as *answers to specific questions* and cannot be detached from those questions. In other words, certain acts of believing count as philosophical once they are properly embodied in the sense sketched out above. These questions do not have to be made explicit each time a philosophical belief is articulated, however they must always be available, so to speak, meaning that they can be extracted and put into words.

Here is a nice illustration of why the latter posit is not trivial. John Searle starts his (1998) book by listing what he calls "default positions" or the Background, that is, beliefs accepted (by everybody, he claims) *prereflectively*; for example, that there is a real world existing independently of our perception or thinking, that we have direct perceptual access to this world, and so on. Searle emphasizes—quite rightly—that these are not *views* proper, thus not *beliefs* proper, since they are so taken for granted that nobody, with the exception of philosophers, questions them. However, even if the Background exists (which, by the way, should be tested empirically by Xphi), why should we assume that all of its belief-like constituents are *philosophical*?

Contrary to what Searle suggests, we can distinguish between different embodiments of *seemingly* identical beliefs (acts of believing) referring to familiar objects' being *real*. Yet only some of these embodiments bring philosophical problems to the table, while, most likely, *all* of them belong to the Background as understood by Searle. Consider this example:

(7) John believes that the chair in his office is real qua answering the question of whether the chair is real or the product of hallucination.
(8) John believes that the chair in his office is real qua answering the question of whether the chair is real or a fake imitation made of paper.
(9) John believes that the chair in his office is real qua answering the question of whether the chair is real or made up of sense-data.
(10) John believes that the chair in his office is real qua answering the question of whether only physical particles are real or whether ordinary things such as chairs are real, too.

Clearly, only (9) and (10) bring in philosophically relevant contrasts (as for 7, hallucination can be thought through by philosophers, exploited in arguments, but there is nothing particularly philosophical about hallucination *itself*). This example is thus supposed to suggest that philosophical beliefs are not just there, lying on the street or flying in the air, waiting for philosophers to articulate and question them; instead, they must be *brought forth* by a specific intellectual pursuit.

It may also be helpful to take a look at the analogy provided by Peter Strawson, who says that "just as the grammarian (…) labours to produce a systematic account of the structure of rules which we effortlessly observe in speaking grammatically, so the philosopher labours to produce a systematic account of the general *conceptual structure* of which our daily practice shows us to have a tacit and unconscious mastery" (Strawson, 1992, p. 7). We used to think of language as a structure whose grammatical rules are embedded in it, and of the said rules as already being there, *in* the language, and perhaps there is some appeal to this claim just as there may be some appeal to the view that there are philosophical problems embedded *in* real-life cases and scientific puzzles. Yet grammar itself, as an abstract scaffolding of language, can become a target of investigation *once it is distilled*, as it were, by some intellectual pursuit. Otherwise there cannot be a properly *grammatical problem*. A particular puzzle functions as a properly grammatical problem or question only when it is recognized as part of this distilled abstract scaffolding. In other words, only when it has the right sort of contrast or embodiment set by other properly grammatical questions and answers. And the same may apply to philosophy.

Someone might argue that the differences between (8) and (9) boil down to differences between various meanings of "real." Perhaps this is also a plausible way of answering the problem of why only some of these statements represent a philosophical puzzle, but I would insist that targeting the problem in terms of contrasts is better for at least one reason, already indicated earlier. Namely, there may be subtle differences between various meanings of "real" and other philosophically pregnant words within philosophy itself, endangering the very possibility of a fruitful philosophical debate, for example, between materialists and phenomenologists, who, most likely, *do* differ when it comes to the content they associate with "real" and "reality." Targeting the issue in terms of questions and contrasts enables us to outline differences between beliefs, while at the same time providing platforms for dialogue insofar as the beliefs are comparable.

Going back to our initial story of Jane, we can now account for the difference between W_n and W_p in yet another way. Namely, at the beginning Jane answers the question pertaining to George's belief/knowledge *as if* the question was of the same kind as, for example, "Do you know that this is the right way or are you just guessing?" asked during a vacation trip, or "Do you know exactly where the keys are or are you going to search the whole place?" put to someone who has forgotten to take their car keys and must retreat from the garage to find them. Answers to these real-life questions can be clearly ascribed to their proper real-life contrasts. Arguably, the question about George in Nichols et al. (2003) *is* different, since it would be difficult to find a real-life contrast for it, but this fact *may not be* apparent and self-evident to a layperson who is supposed to provide an answer in a questionnaire in a flash (as pointed out by Rini, 2015). Jane at this stage simply had no other available contrast for her assessment of George's knowledge; thus, she took the real-life contrast which, arguably, hardly fits there.

However, as Jane acquires more knowledge of philosophical puzzles, she can answer the question about George within a different frame of reference, namely alongside questions such as "Can I know things in themselves or only phenomena?" or "Do I know other minds directly or indirectly?" She becomes an expert *not* because she knows things about Plato or Wittgenstein, but because she is capable, likely due to the mentioned readings (but *not necessarily* due to them), of *placing the question* about George in a completely new contrast that was inaccessible from the everyday standpoint. Now, her further philosophical education results in an augmentation of the "philosophical" contrast, as it were, which leads to a change in her position ("back" to W), yet this change remains within the same frame of reference—within the same broad family of contrasts, as it were. For the sake of brevity, we can articulate the said differences in the following manner:

(11) Jane believes *W* qua answering real-life local knowledge–related puzzles (which is referred to as W_n).
(12) Jane believes *S* qua answering global knowledge–related problems (which is referred to as S_p).
(13) Jane believes *W* qua answering global knowledge–related problems (which is referred to as W_p).

Finally, what makes W_n and W_p distant is the distance between them on *the scale of comparability*.

Now, we should determine whether folk beliefs and proper philosophical beliefs instantiate the extreme scenario in which the two sets are completely incomparable, meaning that their contrast beliefs taken together with their consequences have nothing in common. There is no reason to assume, however, that this must be the case. For at least one thing would be problematic once we take the extreme scenario into account: after all, philosophers are human beings whose awareness of, and interest in, philosophical puzzles arose as part of their dealing with their real-life surroundings, not to mention philosophical problems inspired by developments in science. Drawing a sharp line between the philosophical domain on the one hand and everything else on the other seems way too radical. Until proven otherwise, a more realistic scenario would make the case that there is continuity between everyday beliefs and their contrasts on the one hand, and philosophical beliefs together with their proper contrasts on the other, and that some of the beliefs belonging to these sets are closer to each other while others are more distanced. New challenges appear in this light, but also new prospects for both Xphi and its critics.

To sum up, the primary difference between philosophers and others manifests itself not in the guise of two qua objects based on *subjects*, drawing a sharp line between philosophers and non-philosophers, as the initial proposal would have it, but in the guise of qua objects based on beliefs or *acts of believing*.

Note that whereas it turned out to be hardly plausible to draw a sharp line between philosophers and non-philosophers due to the fact that the latter is an umbrella term standing for a broad variety of different capacities, it is possible to draw a relatively clear line between philosophical *beliefs* (acts of believing) and non-philosophical *beliefs* (acts of believing). The latter line refers to different contrasts (embodiments) attributable to the beliefs in question, thus to different sorts of questions that the beliefs are supposed to answer.

The issue of questions proves crucial for the entire project set forth in this book. In the next chapter, I make the case that at least some questions are spawned by a very peculiar cognitive (intellectual) faculty called erotetic intuition.

2.8 Two Kinds of Commitment

More needs to said about acts of believing, however, because given the variety of relevant qua objects we can have different kinds of believing or different kinds of commitment to a claim. To see things as they stand without any presumption as to how philosophers operate—even though they are our main topic here—we shall take an entirely unrelated profession as an illustrative example. Let us imagine the following conversation:

Person 1: You know, Kazimierz helps people as a firefighter, but when he has a day off, he doesn't care so much about people's lives. He just wants to do the job, but he's personally not passionate about helping others. I think it's bad.

Person 2: Anna wasn't so passionate about helping others either, but recently she changed, and she voluntarily does many good things because she became a firefighter. I mean that once she became a firefighter, she changed her attitude. This job actually changed *her*. And when it comes to Kazimierz, you know, at the end of the day he is way better than Max, who works in the fire department, yet he neglects all his duties, and thus offers no help to others not only as a private person but even as a firefighter.

Now, let us analyze these imaginary cases. Whereas in the first case helping people is ascribed to Kazimierz only *in his capacity* as a fireman and to his actions in this capacity, but not to Kazimierz's actions attributed to him *per se*, helping people is ascribed to Anna's behavior *as such* while the appeal to her role and to her actions in that role is intended to point out what caused the change in her attitude or character. Note that she changed as result of being a firefighter and now helps people whether or not she is actually on duty or even whether or not she plans to continue her career as a firefighter, whereas Kazimierz did not change in the same way, although he keeps acting as he should in his role (he's professional). Then we have Max—he is a peculiar case since he actually plays the role, yet does not act upon it. This is the extreme (theoretically distinguishable) case of an empty or fruitless role: a role that is not enacted in any way (or enacted in an insufficient way).

Therefore, we have two kinds of commitment to the firefighter's duty (setting Max's lack of commitment aside). One is exemplified by Anna. We can say that Anna *qua playing the role of a firefighter* helps people and we can also state that Anna (Anna per se) helps people. The other is exemplified by Kazimierz: he also helps people qua playing the role of a firefighter, but this time we cannot simply skip the qua phrase and say that Kazimierz helps people. For the latter is false outside the strict context of his role.

Now, what does all of this have to do with philosophical beliefs? Keeping in mind the firefighters, I propose to consider an analogous situation involving philosophers. So, this time imagine a philosopher committed to, say, Alfred North Whitehead's metaphysics, who says that the best ontological model of reality describes it as a realm of processes, hence that the thesis "reality consists of processes" is true, meaning that it best reflects current scientific findings. But this does not necessarily mean that the said proponent of Whitehead's view *personally* believes *that reality consists of processes.* The individual in question could say something like: "Well, I am not in a position to make such ultimate personal statements about reality since all I can do is rely on specific models. But as long as science is taken into account, Whitehead's metaphysical model does a good job; it's the one I endorse." Hence, the said person *believes in some specific sense* that reality consists of processes, but at the same time the same person in some other sense does *not* believe that reality consists of processes. Now, imagine that this philosopher has a friend who is not a philosopher but likes to discuss philosophy-related topics. The friend suddenly becomes convinced by the purely theoretical speculations of the Whitehead scholar and personally embraces the belief that reality on a fundamental level does indeed consist of processes.

So, there are at least two kinds of commitment to a philosophical claim that someone can make: one can be called personal, for lack of a better word, while the other is strictly limited to the theoretical context.

The following distinction also captures the idea rolled out above: Endorsing a philosophical thesis does not necessarily mean that one takes it as a *de re* statement influencing one's direct attitude to *things* in the world. A commitment to the philosophical claim may be a *de dicto* pick of the best (in one's current opinion) account of a specific problem within a specific theoretical context. Thus, it is an attitude toward a *proposition, not toward things.* Think of the proponent of Whitehead's metaphysics again. He might say something like "I abstain from holding any *de re* metaphysical *beliefs* about the ultimate substance of reality, but I do hold (*de dicto*) that Whitehead's model is currently the most appropriate given the science we have." The friend, on the other hand, adopts both *de dicto* and *de re* beliefs, yet we can even think of a situation in which *de re* remains that person's firm commitment even when *de dicto* is dropped like a ladder that is not needed any more once he has climbed up to where he wants to be. What initially prompted the personal (*de re*) change in the latter case was just an intellectual game with no proper significance from the non-philosopher's perspective.

We therefore have two kinds of commitment to a proposition, and to a philosophical claim in particular. This has significant ramifications and affects the institutional approach to philosophical competence sketched out in the Introduction. In short, the trouble with the institutional approach is that it is blind to the distinctions just made.

2.9 Against the Institutional Approach

The institutional approach to philosophical competence boils down to the claim—implicitly assumed and implemented in some contexts, or even explicitly articulated—that being a philosopher actually means being a certified specialist in a conventionally defined set of beliefs *considered* philosophical. Certification is one of the goals of institutions like universities, so in order to count as competent in philosophy one needs to have a diploma in philosophy, a related field, or at least to have taken some philosophy courses as part of one's curriculum. And when it comes to the topics deemed philosophical, thus to the assortment of issues that are supposed to be taught, the institutional approach rests on the formal or informal institution of convention—there is nothing essential about any topic that makes it a properly *philosophical* topic, just as there is nothing essential about a work of art that makes it a work of art, as the institutional account of art would have it.

The institutional approach is exemplified by David Bourget and David Chalmers, who conducted an on-line survey that was supposed to elicit the "philosophical views of contemporary professional philosophers" (Bourget & Chalmers, 2014, p. 465). In the same spirit, Nichols et al. (2003) endorse the said approach almost explicitly when they hypothesize that epistemic intuitions depend on how many philosophy courses a person has taken (see Nichols et al., 2003, Hypothesis 3). Still in the same vein, Kuntz and Kuntz (2011) declare that they want to examine "how philosophers conceive of the use of intuition in philosophical methods" (Kuntz & Kuntz, 2011, p. 644), and then they go on to say that the data for their study "were collected via online survey. A survey link was sent to four philosophy list-servers: Philos-L, for philosophy in Europe; Philosop-l, for philosophy in North America (…). The aim was to cast a broad net about *professional* English-speaking philosophers." It turned out that "282 *professional* philosophers agreed to participate" (ibid., p. 646; my highlights). Hence, in order to find a philosopher, one has to make calls to universities, professional associations, and so on.

Moreover, those who challenge the very idea of Xphi (e.g., Williamson, 2011) used to understand philosophical competence or expertise in the same manner, arguing, as Nado (2014) unpacks it, that while the surveys "conducted by experimental philosophers have (…) employed non-philosophers as subjects," the "appropriate evidential source for philosophy is *expert* intuitions—in other words, the intuitions of *professional* philosophers" (Nado, 2014, p. 631; my highlights).

The point here is not to make the case that the institutional approach is plainly wrong. Insofar as it provides an operationalization of the notions of a philosopher and philosophical competence for a specific set of experimental studies, it may be useful and sufficient. It is methodologically fine if one wants to launch an experimental project. However, the point I do stress is that the institutional approach is not enough, not only to warrant Xphi's *philosophical* significance, thus to prove that Xphi in not simply a new branch of cognitive psychology, but, aside from all other interesting coincidences, that it is a new branch *of philosophy* in the first place. Put differently, the institutional approach is not enough to warrant that Xphi's results are indeed about the issues that philosophers have been concerned with "for millennia," as Starmans and Friedman (2012) emphatically stress.

A proponent of the expertise defense might argue that the problem of Xphi's philosophical significance is something that members of the Xphi camp should be worried about, whereas the more traditional philosophers have no reason to care. I disagree. If we don't know how to demonstrate the philosophical significance of Xphi (thus to establish some degree of comparability), we also don't know how to (re)position traditional philosophizing in the age of Xphi and the dynamic development of cognitive science which serves as the role model for the experimental project within philosophy. In other words, provided that traditional philosophers cannot simply ignore Xphi and other empirical studies somehow related to properly philosophical matters, they should aim at finding the proper place for philosophizing in the new context, in the new environment, transformed, whether we like it or not, by the emergence of Xphi and all these other fields. This is why the traditional philosopher must care about how the philosophical significance of Xphi and various other branches of cognitive science can be demonstrated and therefore, also, about the limits of this significance (provided there are some limits, thus that Xphi is not supposed to replace traditional philosophizing).

Going back to Bourget and Chalmers (2014), my main worry, which also applies to the entire institutional approach, is the following: it is actually unclear what kind of commitment to a philosophical view was elicited by their survey. On the one hand, they seem to target the personal, thus *de re*, beliefs of those who count as professional philosophers; on the other hand, however, they hope that since contemporary philosophy is the "leading edge" of the history of philosophy, a better, empirically grounded understanding of today's philosophical views can be of some use to the historians of the future. However, it seems more likely to me that unlike future biographers, these future historians will be, as we are today, interested in *theories* or *claims put forward by today's philosophers*, thus in the abstract theoretical constructions they endorse *de dicto*, rather than their personal beliefs (of course, the philosopher can also adopt a more personal stance, but this will likely be irrelevant to the future historians). And these two things, as we saw, are not the same.

Another question is whether or not the institutional approach provides a viable operationalization of philosophical competence; one that could, for example, carve up a group of people so that future progress in philosophy will emerge out of it. This is the hope that Bourget and Chalmers expressed explicitly, and it likely seemed trivially true to them that the next game-changers in philosophy will come from within the institutional bounds of academic philosophy as we conceive of it today.

However, this is by no means obvious. We can easily imagine a scenario in which a future game-changer in philosophy, a future groundbreaking work, comes from outside the institutional bounds of contemporary philosophy, given the fact that some genuine game-changers at the end of the nineteenth and the beginning of the twentieth centuries were indeed produced by philosophers who originally came from different institutional backgrounds. Consider this: Frege, Whitehead, Husserl, and Tarski all had doctorates in mathematics with no formal diploma in philosophy; Russell was, we may say, in between, formally educated in math as well as philosophy. Wittgenstein had an undergraduate diploma in engineering while his doctorate in philosophy is the subject of popular anecdotes. Carnap's case is quite complicated, too, since he was formally trained as a physicist, yet ended up publishing his dissertation on the concept of space in *Kant Studien* due to certain inconsistencies between the institutional order in academia and the actual dynamics of his work. On the other side of the analytic-continental divide (if one wishes to stick with this strife in the first place), think of Nietzsche, who had a doctorate in philology, or Kierkegaard, who was trained as a theologian.

This sketchy list is not supposed to prove any specific point, but to demonstrate, perhaps, a somewhat trivial point that there is much more to be said about the dynamics of people's philosophizing than the institutional framing allows. Yet, there is non-trivial work to do based on this trivial point if one wants to be seriously prepared for the next game-changers in philosophy, *wherever they may come from*. Although coming up with predictions of this kind would be in vain, the lesson is nevertheless important: we should pay more attention to the *actual dynamics of philosophizing* instead of merely looking at the institutional framework.

2.10 Expertise Defense Revisited

Given that non-philosophical acts of believing are likely de re in most cases, we can hardly confront them with the *de dicto* philosophical claims of professionals. And we must not conflate the *de dicto* philosophical theses articulated in thousands of treatises and books with the *de re* beliefs of their authors. For it seems that only the personal and *de re* beliefs, even those possessed by professional philosophers, can be paired with the beliefs of ordinary people. This brings back the question of the *comparability* of beliefs. For example, if Jane is a philosopher and Kazimierz is not, then we should compare Kazimierz's folk and *de re* beliefs about knowledge or intentionality with Jane's relevant *de re* beliefs, not with her *de dicto* claims spelled out in her papers. However, we would need a tool to elicit the former as distinguished from the latter, and the institutional approach provides no such tool. But perhaps we could imagine experimental settings in which philosophers could enact some *de re* presumptions (so that there would be no obstacle to juxtaposing them with folks) or ones in which non-philosophers could enact some *de dicto* beliefs. By "enacting" I simply mean putting something into practice.

Moreover, one might argue, based on the examples given in the previous paragraph, that it is in principle possible to become a "philosopher," that is, to gain philosophical competence, without an institutional framework, thus perhaps even without necessarily reading Plato or Quine on Plato's beard. To have such competence means to be able to *think philosophically*—to begin exercising one's intellectual capacities in pursuit of philosophizing. If so, it is crucial to dig deeper than the institutional approach enables us to, and shed some light on the nature of this pursuit.

Being aware of all the limitations of the institutional approach, the expertise defense could emphasize the need for more demanding conditions of *what counts* as a proper *philosophical* question, thereby making the case not that laypeople as such have nothing to offer professional philosophy, but that the contrasts usually attributable to folk beliefs make the latter hardly comparable to philosophical beliefs proper. In this rendering, the expertise defense is not about the philosopher vs. non-philosopher (actually postulating strife between them, which is a much too coarse-grained way of looking at the issue, as we saw above), but about the right sort of questions to be addressed by anyone who is contributing to philosophy (regardless of whether that person is a professional philosopher in the institutional sense).

It is a serious issue which does not attract ample attention. Here is how Machery portrays the expertise defense before providing his counterarguments in favor of Xphi:

> Studies conducted by experimental philosophers do not examine [according to the expertise defense] the judgments made by *the right kind of people*. In particular, they examine the judgments of people without the pertinent knowledge or skills. (Machery, 2017, p. 155; italics mine)

Perhaps this is the intention of some expertise defenders, but it is not the only way of framing the argument. Speaking of the "right kind of people" can be misleading. The point, in my understanding, is not that some people are in a fixed state of having no idea what philosophy is about, as if philosophy was some sort of mysterious insight. A more realistic approach involving conditions of comparability is that Xphi studies must not be done on people who happen to lack the right kind of contrasts, thus the right kind of questions that would help them frame the scenario they are asked to respond to.

However, both sides of the dispute should be blamed for this, for the expertise defense, as it is usually articulated, misses the fact that some of the views expressed by folks likely *could* be compared to the views of professional philosophers once it is proven that these folks are employing properly philosophical contrasts. For there is no a priori reason to deny that some of the folks may indeed be successfully provided with the right sorts of contrasts.

At the other extreme, experimentalists neglect the fact that comparability must be *achieved, not assumed*, which would most likely require the advancement of Xphi methods (see Andow, 2016; as for methodological issues, Xphi is becoming increasingly self-aware; see the most recent collection by Fischer & Curtis, 2019). Laypeople would have to be subjected to certain tasks or training so that their beliefs could move on the comparability scale closer to "philosophical" beliefs. For example, more thorough reflection by the participants of the experiments done by Nichols et al. (2003) could perhaps render folk intuitions better suited to solve the properly philosophical puzzle that Putnam (1981) originally envisioned.

This brings us to another defense, correlated, but not identical, to the expertise defense, namely the strategy referenced as the "reflection defense" by Kneer et al. (2020). Simply put, this defense says that philosophers are skilled in detailed, analytic thinking, and therefore presenting folks with short vignettes in an environment in which they are not instructed but nevertheless prompted to give intuitive answers can be of no significance to the analytic work of the professional philosopher. Now, Kneer et al. (2020) as well as Machery (2017) answer this challenge, yet in a way that is by no means satisfactory from the standpoint of the reflection defender (it may be satisfactory for other purposes which are not discussed here). For the word "reflection" is somewhat ambiguous, hindering prospects of an operationalization that both sides of the conflict would agree upon. As a result, the Xphi people operationalize reflection in a way that is accepted in the field of cognitive psychology. For example, as summarized by Machery (2017), they recommend that participants read the vignette several times, thus giving them more time, as if time were the issue. They don't tell them, however, what to do with this time, what reflective strategy to employ, and so on. They also provide incentives such as money: the participants are to be rewarded only if they provide the "right" answer, which is supposed to make them more reflective and scrupulous. But how so? There is no way to distinguish those who actually plunge into a philosophical-like way of thinking from those who don't.

A more demanding procedure would require that participants do more than just read the vignette. It might perhaps involve discussing the story in greater detail in focus groups. And this in turn would require the experimenters to do more than just hand out vignettes with questions, as moderating group discussion, for example, would make it necessary to reform the Xphi methodology (Andow, 2016).

In response to all this, a proponent of Xphi, at least one who is reluctant with regard to any reform and finds my arguments unconvincing, might come up with two points. First—that the right sort of contrast is delivered to the folk precisely by the vignette, for example, by Omar's story, and then by the exact question the participant is expected to answer. Otherwise, if the proper contrast needed to be introduced with more precision, one would risk suggesting a particular answer.

My reply to this remark reflects the more general perspective adopted in this book. Let me begin by noting that Xphi kicked off as a sharp critique of "armchair" philosophy, thus of a certain representation (perhaps misrepresentation) of the way philosophy has traditionally been done, allegedly boiling down to purely a priori theorizing, likely based on private intuitions, generalized without any justification. It is not the purpose of this chapter to assess whether this characterization is fair. The point is different, namely that while Xphi is set against "armchair" philosophizing, *it places its own participants* in the armchair, so to speak, and asks them to consider things they have never reflected upon before. The Xphi experimenter might reply that this is exactly what traditional philosophers do, but that is plainly wrong. We can debate the (im)plausibility of various philosophical methods, but it is obvious that philosophers bring to their armchair a significant bulk of knowledge from other disciplines, even if they don't refer to it explicitly, having filtered it through their philosophical interests. They also bring their everyday experiences, filtered in a similar way, so that in the end they sit down in an armchair loaded with materials to be theoretically digested (whether this digestion produces anything of value is another issue). Meanwhile, the folk participants of an Xphi study are really *decoupled* from their everyday experiences or topics, and suddenly subjected to questionnaires depicting either completely unfamiliar stories like the one about Omar, or seemingly familiar cases, trivial at first glance, such as the Gettier scenarios. The point is, therefore, that one vignette and one question are hardly sufficient to *introduce* the folk participant to the right sort of contrasts, thus to the right sort of questions which provide the background that Putnam's story or Gettier's scenarios should be seen against. Introducing such contrasts, thus the proper context in which these thought experiments make sense, would require more from the experimenters—not only to discard their philosophical armchairs, but also to allow the participants to leave *their* armchairs, too. That is why I move step-wise in these considerations toward a more embodied and enactive approach to cognition in general, and to thinking, including philosophizing, in particular.

As for the methodological remark pointing to the risk of suggesting an answer to the participant, it hints at a much broader debate on whether qualitative methods can deliver scientifically viable results. There is no room here to go into the details of this issue. I assume that the said risk can be managed and kept under control (see e.g. Smith, 2011). After all, this has been done successfully for a couple of decades now, for example in health psychology.

Secondly, the Xphi man might also point to the fact that the expression "the right sort of questions," referring to the proper philosophical contrast, is vague. Which questions are *the right ones* that distinguish properly philosophical contrasts from non-philosophical ones? It is unclear, especially once we abandon the institutional approach. Indeed, there is likely no ready-made and unproblematic answer to this puzzle.

So, while Chaps. 3 and 5 take on the problem of where questions in general come from and what cognitive capacity gives rise to them, Chap. 4 seeks to specifically characterize philosophical questions.

2.11 Conclusion

A general lesson that I take from the above considerations is that we should refocus our attention from beliefs as separate "atoms" existing on their own, to *question—belief (answer)* pairs. For without questions, explicitly articulated or implicitly embedded in a particular context, it is virtually impossible to figure out what a given belief is about and what other beliefs it can be compared with. Given the abundant literature on beliefs, what this means in practice is that we should pay more attention to questions and their origin. As for the latter, we also need to pay more attention to the specific circumstances in which questions arise, thus, to their role in the actual functioning of a cognitive agent. All this includes philosophy, philosophical beliefs, philosophical questions, and—as I prefer to say—*acts of philosophizing*.

Notes

1. Let "$x\backslash\varphi$" stand for "x qua φ," as Fine (manuscript) proposes. We have:

EXISTENCE: A qua object $x\backslash\varphi$ exists if and only if x exists and possesses φ.
IDENTITY: Two qua objects are the same iff their bases and glosses are the same.
INHERITANCE: Given that a qua object exists, it will have exactly the ordinary properties of its basis.
PART: Both x and φ are parts of $x\backslash\varphi$.
FOUNDATION: There is no sequence of objects x_1, x_2, x_3, …, with x_2 the basis for x_1, x_3 the basis for x_2, and so on ad infinitum. (Fine, *Acts and Embodiments*, manuscript, p. 5)

2. Someone might argue that mental acts are not good candidates for bases since they are metaphysically problematic; instead, we should speak of certain physio-chemical processes in our brains which *happen to be* acts of believing or something else. Arguably, it is possible that we have to go down to biological functioning in search of ultimate bases, yet it can hardly be only about the brain and the nerves; it must also be about the interactions between the whole organism and its environment. I wish to stay neutral concerning this issue, however, and, as for now provisionally, let me stay with mental acts as bases.

References

Andow, J. (2016). Qualitative Tools and Experimental Philosophy. *Philosophical Psychology, 29*(8), 1129–1141.

Bäck, A. (1996). *On Reduplication. Logical Theories of Qualification*. E. J. Brill.

Baz, A. (2017). *The Crisis of Method in Contemporary Analytic Philosophy*. Oxford University Press.

Beebe, J. R. (2012). Experimental Epistemology. In A. Cullison (Ed.), *Companion to Epistemology* (pp. 248–269). Continuum.

Bourget, D., & Chalmers, D. J. (2014). What Do Philosophers Believe? *Philosophical Studies, 170*(3), 465–500.

Buckwalter, W. (2010). Knowledge Isn't Closed on Saturday: A Study in Ordinary Language. *Review of Philosophy and Psychology, 1*(3), 279–296.

Buckwalter, W. (2012). Non-Traditional Factors in Judgments about Knowledge. *Philosophy Compass, 7*(4), 278–289.

Buckwalter, W., & Schaffer, J. (2015). Knowledge, Stales, and Mistakes. *Noûs, 49*(2), 201–234.

Egler, M., & Ross, L. D. (2018). Philosophical Expertise Under the Microscope. *Synthese*, preprint https://doi.org/10.1007/s11229-018-1757-0

Fine, K. (1982). Acts, Events, and Things. In W. Leinfellner, E. Kraemer, & J. Schank (Eds.), *Language and Ontology. Proceedings of 6th International Wittgenstein Symposium* (pp. 97–105). Hölder-Pichler-Tempsky.

Fine, K. (2018) Acts and Embodiment, unpublished manustript available online: https://www.academia.edu/35032853/acts_and_embodiment_pdf [accessed on 09.21.2021]

Fischer, E., & Curtis, M. (Eds.). (2019). *Methodological Advances in Experimental Philosophy*. Bloomsbury Academic.

Gettier, E. (1963). Is Justified True Belief Knowledge? *Analysis, 23*, 121–123.
Hamblin, C. (1958). Questions. *The Australasian Journal of Philosophy, 36*, 159–168.
Harrah, D. (2002). The Logic of Questions. In D. M. Gabbay & F. Guenthner (Eds.), *Handbook of Philosophical Logic* (Vol. 8, 2nd ed., pp. 1–60). Springer.
Hiż, H. (1962). Questions and Answers. *Journal of Philosophy, 59*(10), 253–265.
Kauppinen, A. (2007). The Rise and Fall of Experimental Philosophy. *Philosophical Explorations, 10*, 95–118.
Kneer, M., Colaço, D., Alexander, J., & Machery, E. (2020). On Second Thought: Reflections on the Reflection Defense. Forthcoming in *Oxford Studies in Experimental Philosophy*. https://www.researchgate.net/publication/325465989_On_second_thought_Reflections_on_the_reflection_defense
Knobe, J. (2003). Intentional Action and Side Effects in Ordinary Language. *Analysis, 63*(279), 190–194.
Kuntz, J. R., & Kuntz, J. R. C. (2011). Surveying Philosophers About Philosophical Intuition. *Review of Philosophy and Psychology, 2*, 643–665.
Ludwig, K. (2007). The Epistemology of Thought Experiments: First Person Versus Third Person Approaches. *Midwest Studies in Philosophy, 31*(1), 128–159.
Machery, E. (2017). *Philosophy Within Its Proper Bounds*. Oxford University Press.
Machery, E., Stich, S., Rose, D., Chatterjee, A., Karasawa, K., Struchiner, N., Sirker, S., Usui, N., & Hashimoto, T. (2015). Gettier Across Cultures. *Noûs, 51*(3), 645–664.
Merleau-Ponty, M. (1962). *Phenomenology of Perception* (C. Smith, Trans.). Kegan Paul.
Nado, J. (2014). Philosophical Expertise. *Philosophy Compass, 9*, 631–641.
Nichols, S., Stich, S., & Weinberg, J. (2003). Metaskepticism: Meditations in Ethno-Epistemology. In S. Luper (Ed.), *The Skeptics* (pp. 227–247). Ashgate Press. Quotations from S. Stich, *Collected Papers. Vol. 2. Knowledge, Rationality and Morality, 1978–2010*. Oxford University Press, 2011.
Poli, R. (1999). Qua-Theories. In L. Albertazzi (Ed.), *Shapes of Forms* (pp. 245–256). Kluwer.
Putnam, H. (1981). Brain in a Vat. In H. Putnam (Ed.), *Reason, Truth, and History* (pp. 1–21). Cambridge University Press.
Putnam, H. (1987). *The Many Faces of Realism*. Open Court.
Rini, R. A. (2015). How Not to Test for Philosophical Expertise. *Synthese, 192*, 431–452.

Roberts, P., Andow, J., & Schmitdtke, K. A. (2017). Lay intuitions about Epistemic Normativity. *Synthese*. https://doi.org/10.1007/s11229-017-1371-6
Ryle, G. (1949). *The Concept of Mind*. University of Chicago Press.
Schaffer, J. (2007). Knowing the Answer. *Philosophy and Phenomenological Research, LXXV*(2), 383–403.
Schwitzgebel, E. (2009). Do Ethicists Steal More Books? *Philosophical Psychology, 22*, 711–725.
Schwitzgebel, E., & Cushman, F. (2012). Expertise in Moral Reasoning? Order Effects on Moral Judgment in Professional Philosophers and non-Philosophers. *Mind & Language, 27*, 135–153.
Schwitzgebel, E., & Cushman, F. (2015). Philosophers' Biased Judgments Persist Despite Training, Expertise and Reflection. *Cognition, 141*, 127–137.
Schwitzgebel, E., & Rust, J. (2016). The Moral Behaviour of Ethicists. In J. Sytsma & W. Buckwalter (Eds.), *A Companion to Experimental Philosophy* (pp. 225–233). Wiley-Blackwell.
Searle, J. R. (1998). *Mind, Language and Society: Philosophy in the Real World*. Basic Books.
Smith, J. A. (2011). Evaluating the Contribution of Interpretative Phenomenological Analysis. *Health Psychology Review, 5*, 9–27.
Starmans, C., & Friedman, O. (2012). The Folk Conception on Knowledge. *Cognition, 124*, 272–283.
Strawson, P. (1992). *Analysis and Metaphysics. An Introduction to Philosophy*. Oxford University Press.
Tobia, K., Buckwalter, W., & Stich, S. (2013). Moral Intuitions: Are Philosophers Experts? *Philosophical Psychology, 26*(5), 629–638.
Twardowski, K. (1911/1999). O czynnościach i wytworach. Kilka uwag z pogranicza psychologii, gramatyki i logiki. Originally Published in 1911; Translation available in: K. Twardowski, *On Actions, Products and Other Topics in Philosophy*. Rodopi.
Werner, K. (2020). The Little Word 'as'. On Making Contexts and Aspects Explicit. *Axiomathes, 30*, 69–90.
Williamson, T. (2007). *The Philosophy of Philosophy*. Blackwell Publishing.
Williamson, T. (2011). Philosophical Expertise and the Burden of Proof. *Metaphilosophy, 42*, 215–229.
Wiśniewski, A. (1995). *The Posing of Questions. Logical Foundations of Erotetic Inferences*. Springer.
Woolfolk, L. R. (2013). Experimental Philosophy: A Methodological Critique. *Metaphilosophy, 44*(1–2), 79–87.

3

Erotetic Intuition: Toward a Logic of Questions, and Beyond

3.1 Questioning Questions

How do we manage to ask a question? How do we manage to combine questions and answers? It turns out that these puzzles are by no means trivial. In this chapter, I shall propose that a specific intellectual capacity—*erotetic intuition*—is responsible for problematizing a given fact or entity, that is, contrasting it with a more or less specified set of alternative scenarios (so that the subject knows her epistemic desiderata or information requests). Crucially, I will show that the machinery of erotetic intuition cannot be fully articulated in propositional terms. My argument, like the one put forward by Elijah Chudnoff (2011) in favor of the perceptualist account of intuition, draws on an analogy with a similar debate going on in the philosophy of perception—the debate on whether or not the content of perception is propositional. Hence, although Chudnoff makes no reference to the issue of questions, in some sense I line up with him and argue that erotetic intuition is a kind of experience in which certain contents are provided to us. To put it as simply as possible: it is the capacity to see or recognize problems to be tackled in one's surroundings.

K. Werner, *The Embodied Philosopher*, New Directions in Philosophy and Cognitive Science, https://doi.org/10.1007/978-3-030-79964-9_3

It must also be stressed, however, that speaking of erotetic *intuition* is here a kind of via negativa, almost like in negative theology, for at the end of the day the word "intuition" covers our ignorance and a yawning gap in our understanding of cognition that we are hardly capable of overcoming at this point. Namely, we have no comprehensive account of how it is possible for a subject capable of cognition not just to process information from her environment, but also to spontaneously problematize this or that fragment of the said environment, thereby not just demanding *more* information about an entity that she has already problematized, but bringing something new to the domain of questioning, so that we expect to learn *anything* about it in the first place.

Rolling out the notion of erotetic intuition in greater detail requires a short introduction to the problem of intuition, as well as at least a brief glimpse at the discipline called logic of questions, erotetic logic or erotetics. I will begin with the latter, then move on to the former, and finally combine the two, making the case for erotetic intuition. Once this is done, a specific problem connected to the proposed concept emerges. Namely, it is necessary to explain the postulated *directness* of the intuitive erotetic grasp of things.

3.2 On Posing Questions: Preliminary Remarks on Erotetics

Questions have been studied by philosophers since the time of Aristotle. The art of questioning, if such a phrase can be used, has also been at the heart of several views regarding how philosophy should (or should not) be pursued, starting from the Socratic method of questioning which aimed at eliciting what was already tacitly known to the respondent; through Descartes' meditations, where the philosopher questions himself as to what piece of knowledge is firm enough to withstand doubt; up to our own times when the art of philosophical questioning was scrutinized by Roman Ingarden (1925) under the heading of "essential questions" (e.g., "what is *x*?") and severely criticized in logical analysis by Wittgenstein (1922) and Carnap (1950).

However, despite having a long history, a disciplined study of questions as part of modern formal logic is a relatively new pursuit only going back to the 1950s and 1960s, with the publication of Prior and Prior's (1955) and Hamblin's (1958) papers followed by Hiż (1962), Åqvist (1969, 1971), Kubiński (1967, 1980), Belnap and Steel (1976), Szaniawski (1973), Hintikka (1975, 1992), Wiśniewski (1994, 1995), and Harrah (1985), among others (see Harrah, 2002 for an overview).

Charles L. Hamblin (1958) formulates the following postulates:

P1. An answer to a question is a statement.
P2. Knowing what counts as an answer is equivalent to knowing the question.
P3. The possible answers to a question are an exhaustive set of mutually exclusive possibilities.

I shall adhere to these points, but with one restriction, pertaining to P2. Namely, the statement rightly brings our attention to the close connection between questions and answers: understanding a question necessarily involves understanding the proper list of statements representing possible answers. But it also works other way round, which is perhaps not what Hamblin had in mind. Namely, understanding a statement which happens to be an answer to a fixed question necessarily involves understanding the question. Yet the notion of equivalence is perhaps too strong, as it hints at the possibility of reducing questions, that is, spelling them out in terms of the set of statements answering it. Note, however, that it is one thing to *represent* a question by providing a list of possible answers to it, and another to *reduce* the former to the latter.

David Harrah (1987) distinguishes the "abstract interrogative" approach to studying questions, adopted, among others, by Belnap (1983), in which questions are identified with sets of alternative answers, but do not actually say or command anything. In other words, to refer to a question it is sufficient, from the standpoint of logic, to refer to the set of sentences stipulated as answers to that question, without the need to refer to any epistemic or volitive factors on the questioner's side. I am not in a position here to decide to what extent this approach works well in erotetic logic. Yet for our purposes, I am strongly aligned with another approach, pioneered by Åqvist and Hintikka, which merges erotetic logic with epistemic logic, taking the epistemic position of the questioner seriously. In this approach, a question expresses a certain demand or request—a *demand for information*. Therefore, questions are expressions that need to be analyzed on their own, not as mere indicators of something else. They articulate a certain epistemic state of the questioner.

Therefore, Hintikka rightly stresses that we should distinguish between the pure logic of questions and the epistemic logic of questions. While the former takes questions simply as expressions in a given language, somehow built into the logical machinery of that language, the latter takes questions as part of someone's pursuit of knowledge. Here I take the epistemological perspective, drawing partly on Hintikka's own proposal.

Now, P3 brings up the most pregnant issue: speaking of answers requires some notion of a possible world or possible scenario. What this means is that each answer is a member of a wider group of relevant alternatives—alternative ways of satisfying the demand for information. If I ask who killed John F. Kennedy, the relevant alternatives include all possible courses of events with various possible killers, and the task is to nail down which of these scenarios actually happened. Hintikka's erotetics spelled out in terms of relevant alternatives is likely the source of Schaffer's (2007) concept of contrastive knowledge discussed in Chap 3.

3.3 Questions, Desiderata, and Knowledge

There are several ways in which we can distinguish various kinds of questions. According to Hintikka (1975), a line should first be drawn between propositional questions and *wh*-questions, including *who-*, *what-*, *which-*, *when-*, and *where*-questions. This demarcation is determined by one factor, namely what the set of relevant alternatives consists of. The relevant alternatives in the case of a propositional question, which may also be dubbed a *whether*-question, consist, as the name itself suggests, of propositions. For example, the question of whether *Oswald killed president Kennedy* or *Jones killed president Kennedy*, or *the unknown second shooter killed president Kennedy* contains three statements spelling out three possibilities taken into account by the questioner. By contrast, in the case of a *wh*-question, the relevant alternatives consist of the values of a bound variable, thus of objects, not statements. For example, when one asks "who lives in this house" this means that there is *x* such that *x* lives in the house and one wants to know who *x* is. The set of relevant alternatives contains all those who might live in the said house, thus all those who might turn out to be the value of the bounded variable *x*. If the question is "who killed president Kennedy?" the set of relevant alternatives consists of all potential killers.[1]

As already flagged, questions are expressions of a certain epistemic state of the questioner, that is, of a certain demand or request taking the form "Let it be the case that I know…" (Åqvist, 1971) or "bring it about that I know…" (Hintikka, 1975). This is the question's desideratum: it indicates what the questioner wishes to know if the true answer is provided. In some cases, as indicated by Lewis and Lewis (1975), it may also be something like "Make it the case that you tell me truly…," where it is not knowledge that the questioner demands, but rather a confirmation of what he already knows.

Therefore, if I ask, for example, "who killed president Kennedy?" the desideratum assumes, firstly, that some *x* killed Kennedy, where the variable *x* ranges over persons, and the desideratum can be read as "bring it about that I know the individual who satisfies the condition '*x* killed president Kennedy.'" Simply put, "bring it about that I know the one who killed Kennedy." If the question is "Did Oswald or Jones kill Kennedy?" the desideratum is "bring it about that I know whether Oswald killed Kennedy or Jones killed Kennedy," which means "bring it about *that I know that* Oswald killed Kennedy *or that I know that* Jones killed Kennedy." So there are two kinds of desiderata, reflecting the two kinds of questions spelled out above. In propositional questions a desideratum takes the form of an alternative containing propositions that describe possible states of affairs or courses of action, while in *wh*-questions desiderata pertain to objects or entities that are supposed to become known.

It is a serious concern to any system of erotetic logic whether it is rich enough to encompass all (at least all actually used) kinds of questions that appear in language. It is not a particular concern for us here, however. This chapter seeks to solve a more fundamental puzzle, namely *how it happens that a question emerges* in the first place. What is so specific about the erotetic bond, so to speak, between the questioner and his surroundings. By all means, this is a deeply non-trivial puzzle. After all, as already flagged in the previous chapter, questions are not merely there, in the outside world, simply waiting to be picked up. They result from the problematization of a selected part or aspect of the world. But *how is it possible to problematize anything*?

Given the puzzle we want to tackle in this chapter, it is not so important to make detailed distinctions between the various linguistic forms questions take. Instead, I shall draw on the notion of desideratum, which is at the core of every erotetic pursuit, and therefore the purpose will not be to provide a high-resolution representation of all possible forms of question, but to unify this variety at the cost of high resolution. This will nonetheless enable us to see the bigger picture, thus to pin down what characterizes the erotetic situation *as such*.

That said, my short argument aims to show that while the distinction between propositional *whether* questions and *wh*-questions is clear and valid, it nonetheless pertains to the level of linguistic manifestations of desiderata, reflecting first of all *how informed* the questioner is at the moment of asking the question. It is not a fundamental distinction referring to two genuinely distinct epistemic states of the questioner. At the same time, Hintikka's distinction does hint at a more fundamental issue. Namely, whereas in *wh*-questions the actual course of actions is assumed, meaning that we know what happened (e.g., the 35th president of the United States got killed) and are seeking more information about how and why this occurred, *whether*-questions may serve to problematize the fact that the actual scenario is undecided. This is because some questions do not assume upfront which of the possible worlds is the actual one, but instead express the need to determine this.

For the sake of the argument, consider the example of a police officer who asks *whether Oswald or Jones killed John F. Kennedy*. This interrogation reflects the officer's present knowledge that there are only two alternative scenarios. Each of the scenarios is articulated in the form of a proposition and "whether" indicates the relevant alternative that the detective faces. On the other hand, when the question is "who killed president Kennedy?" the detective may have no suspects whatsoever. The question thus reflects his or her ignorance, and therefore the list of relevant alternatives consists of all people, most of them unknown at the moment the question is posed, who were present at the crime scene in Dallas. From the detective's perspective, the pregnant difference between these two questions is that in the case of "who killed president Kennedy?" they may know nothing, to the effect that they must seek the killer among possibly hundreds of people. So, the variable in the desideratum "bring it about that I know *x* who killed Kennedy" ranges over all of these individuals, whereas in the Oswald vs. Jones case the police knows a great deal and must only nail down whether Oswald or Jones is the killer. This knowledge is reflected in the *expression* of their desideratum, as they can significantly limit the range of possible scenarios down to two: the scenario containing Oswald and the scenario containing Jones. Therefore, the detective is not forced to make use of a variable ranging over a large group of unknown individuals, but can instead use the more convenient *whether* template. Yet, *a priori*, if the police had a list of all the people who were present at the crime scene, they could also express their request for information without the use of variables, in the form of an enormously long *whether*-question. It would be something like: "Bring it about that I know that Jones killed Kennedy or I know that Smith killed Kennedy, or I know that Kowalski killed Kennedy….etc." Clearly, the *whether* template is utterly impractical in this context, but not impossible. Therefore, if reeling through all possible scenarios is practically impossible or useless, the easiest way to express a demand for information in a situation of initial ignorance is to ask "who/what did so and so?" thereby making use of a specific logical tool—a variable running over all individuals. I therefore claim that what is really at stake in the distinction between *whether*-questions and *wh*-questions from the epistemic angle is the level of initial knowledge, and thus the relevant level of ignorance, on the side of the questioner. Both forms can *in principle* express the same desideratum.

3.4 Nicholas Rescher's Approach

Nicholas Rescher (2000) comes up with a different classification of questions and I shall bring it up here because it enables us to abstract from Hintikka's distinction between *whether*- and *wh*-questions. For Rescher there is a certain *condition* written into the content of any question, and questions vary when they take different stances toward this condition. Unfortunately, he gives no examples, but we may most likely take, for instance, a statement of fact such as "president Kennedy was killed" as a condition with respect to which we can ask, for example, *why* this condition is realized, meaning why Kennedy was killed; *when* and *where* the condition was realized, meaning when and where Kennedy was killed, and so on. The whole list includes the following for a fixed condition C:

> *Why* questions (Why is condition C realized?)
> *When* questions (When is condition C realized?)
> *Where* questions (Where is condition C realized?)
> *How* questions (How does condition C come to be realized?)
> *Is-it-true* questions (Is condition C realized?)
> *Is-it-possible* questions (Can condition C be realized?)
> *What if* questions (What ensues if condition C is realized?)
> and so on. (Rescher, 2000, p. 1)

Rescher's notion of *condition* is very abstract; thus, when he unpacks his classification in greater detail it turns out that "condition" may refer to "the world's occurrences," "facts about the world," "situations in nature," "space-time placement with regard to the condition that is in question," "natural processes," or even, at the most abstract level, "what is possible" as distinguished from what is not possible or "imaginable possibilities" (see Rescher, 2000, pp. 2–3; points 1–6). Simplifying this rich range of references, I propose that in a question of any kind, condition C describes some state of affairs or a course of action in the world—an actualized one or a possible one. For example, when one asks "who killed president Kennedy?" the condition refers to the course of action that took place in Dallas that day and *problematizes* a certain aspect of it, namely the cause of the course of action (in this case, the killer). Interestingly, it seems that from Rescher's standpoint there is no separate kind of questions such as *what*- or *who*-questions, nor is there any need to distinguish *whether*-questions. I shall comment on this later on.

3.5 Expanding Rescher's Approach: Desiderata and Possible Worlds

Someone might complain that Rescher ignores the previously stressed issue of relevant alternatives. Note, however, that alternatives can easily be added to Rescher's schema so that, for example, we obtain the *why* template "why is *C*1 the case *and not C*2, *C*3,…?"; for *how* questions we obtain "how does condition *C*1 come to be realized *instead of* *C*2, *C*3…?"; for *is-it-true* questions we obtain "is condition *C*1 realized *or* *C*2, *C*3…?" Another interesting option is the contrastive version of *what-if* questions in which we can consider the ramifications of one scenario's coming into existence against the background of alternative scenarios: "what does happen if condition *C*1 is realized instead of *C*2, *C*3…?"

However, my point is actually not that we can *add* relevant alternatives to Rescher's list. Instead, I am prone to argue that they are *already there*, yet implicitly, meaning that Rescher does not make them an issue. Nevertheless, they must be there, otherwise certain questions would not be genuine questions. In other words, I argue that it is essential for a given expression's being *a* question that it contrasts one condition with a different condition. For example, when I ask "God, why did this terrible accident happen to me?" I obviously contrast the actual scenario containing the accident with there being no accident, thus with a variety of unspecified scenarios not involving the accident. To take another example, what sense could it make to ask, say, "what if I drop out of school and set out on a year-long journey around the world" without contrasting the ramifications (the "what-if" part) of this out-of-the-box scenario with the ramifications of a more standard course of action in which I decide to continue studying until graduation? Finally, when someone asks "How did it come about that World War II broke out in 1939?" the very act of questioning—provided that it is not an artificial situation such as an exam but a genuine act of pondering—seems to be driven by one's imagining an alternative scenario in which the world powers of that period somehow come to the conclusion that confrontation would not be in their interest.

Based on this reading of Rescher and my previous considerations on Hintikka's proposal, I propose the following general understanding of a desideratum (I continue numeration from Chap. 2):

(14) General desideratum: bring it about that I know condition C_1 *as contrasted with* a (fully or partially specified) set of alternative conditions $C_2,\ldots,C_n$.

Here is the central idea: it is constitutive of any desideratum, thus of any request for information/knowledge, that it contrasts the condition that is problematized with a variety of (possible) alternative scenarios. Actually, problematization as such boils down to recognizing a given fact against the background of alternative states of affairs.

Now, it seems that each kind of questions distinguished by Rescher has its own unique characteristics in terms of its desideratum. We have:

(15) *General "how-come"* desideratum: Bring it about that I know the causes of condition C_1 as contrasted with possible causes of $C_2,\ldots,C_n$.

(16) *General "what-if"* desideratum: Bring it about that I know the effects of C_1 as contrasted with possible effects of $C_2,\ldots,C_n$.

Both of these *general* forms of desiderata involve causal relation and contrast the actual world containing the actual condition lettered C_1 with a set of possible alternative conditions. Further on, we have, inter alia:

(17) *General "why"* desideratum: Bring it about that I know the reasons (e.g., practical or moral) for C_1 as contrasted with $C_2,\ldots,C_n$.

The "is-it-true" schema is different and I shall focus on it briefly later on.

3.6 *Desideratum–Saturata* Pairs

Note that all the desiderata articulated so far are undetermined: when we know someone's desideratum, we can have a general idea of what the questioner seeks to know, but when it comes to specifics, it may be the case that we still need to utter "Please specify what you want to know," "What do you mean exactly?" or the like. And it becomes clear what *exactly* it is that the request for information contains once the questioner provides us with at least a smattering of possible answers. For example, if the desideratum is "bring it about that I know how it came that World War II broke out in 1939," what the questioner specifically wants to know becomes clear once they are, say, dissatisfied by an answer referring to Hitler's ideology, and retort something like "Yes, ideology drove him, too, but I am not asking now about the ideological background of the war, but rather about the geopolitical facts and trends that led to the outbreak of WWII." In this context, possible answers stating that Germany wanted to gain access to the Soviet Union's natural resources or that Germany wanted to control the flow of goods arriving in Europe from overseas (which is why it invaded the Netherlands); and so on, would *better serve the epistemic needs* of the questioner.

Therefore, I propose that for each desideratum, there is also a saturatum, to use the same convention based on Latin (from the Latin "saturare" meaning "to fill" or "to satisfy"), which makes it clear and explicit what the request for information is about. We can define several kinds of *desideratum–saturata* pairs. For example:

(18) *General "how-come" desideratum–saturata pair*: Bring it about that I know S_1 that is the cause of condition C_1 as contrasted with possible causes of $C_2,\ldots,C_n$; *or* I know S_2 that is the cause of condition C_1 as contrasted with possible causes of $C_2,\ldots,C_n$, *and so on.*

Respectively for other kinds of desiderata.

All that being said, it turns out that the capacity to imagine alternative possible worlds is inherent in both questions and answers. Therefore, we can speak of *two* sets of possible worlds: firstly, in the act of questioning or problematizing itself a fixed entity or course of action is mentally placed or intellectually recognized *against the background* of alternative events or courses of action; and then, secondly, there comes another set of possible scenarios, one of which is supposed to answer the question. As for the example given, if one asks "how did it come about that World War II broke out in 1939?" the actual course of events, that is, the one that

Table 3.1 "How did it come about that World War II broke out in 1939?" Examples of the two groups of alternatives

Alternatives (possible worlds) that *constitute the problem*: the (problematized) actual course of events is highlighted	Alternatives (possible worlds) delivering answers; that is, possible explanations of why the highlighted scenario came into existence.
1. Germany attacks Poland in September 1939 and World War II begins.	**A.** Hitler attacks Poland in order to open up a path to attack the Soviet Union.
2. Germany peacefully resolves its conflict with Poland and there is no war.	**B.** Hitler is determined to attack Poland driven by his ideology of the *Übermensch*.
3. France, aligned with Poland, launches an all-out attack on Germany's west flank, thereby ending the war within a short time (thus no world war either).	**C.** Hitler has a nightmare the night before September 1, 1939, in which Polish seventeenth-century hussars kill him, and, driven by his superstitions, he decides to attack Poland.

unfortunately took place (the war), is contrasted with alternative scenarios in which, to take just two options from a large pool, Hitler did not attack Poland or France launched a more all-out attack from the West flank in September 1939, thereby ending the conflict within a short time. These alternatives constitute *the problem* one wishes to tackle. Then, there is another group of alternative options *answering* the problem, thus eliciting the likely causes of the actual course of events (the war) in contrast to those only imagined (no war). Table 3.1 summarizes these options (it is obviously just an illustration). The alternatives 2 and 3 tell us why 1, thus the realized scenario, is *a problem*. It is a problem, thus *something we need to ask about*, because 2 or 3 *could also have happened*. The question is, therefore, why 1 actually happened instead of 2 or 3. The alternatives A–C are possible explanations of *that* fact.

The take for us is that thinking in terms of possible worlds, thus the capacity to imagine alternative scenarios, is the crux of our erotetic capacity—the capacity to come up with a *desideratum–saturata* pair. I shall develop this simple idea.

The *desideratum–saturata* pair mirrors the *question–answers* pair posited by Hiż (1962), briefly outlined in the previous chapter. Someone might therefore argue that it brings nothing new to the conversation, and hence that there is no need to further complicate the conceptual framework rolled out in this chapter. Yet, there is one reason to introduce the alleged complication: note that when it comes to the *question–answers* pair, it is assumed that the sought knowledge, thus a collection of answers, is propositional. Questions, therefore, are also considered as resulting from a certain intellectual processing of propositions, as when one moves from "there is *x* such that *x* killed president Kennedy" to "I want to know that *a* killed Kennedy or that *b* killed Kennedy, etc."

Meanwhile, it is not necessary to limit the range of desiderata to those pertaining to propositional knowledge, nor to narrow down possible *saturata* to propositional structures. This means that, in principle, one's epistemic needs can manifest in a different form, especially as action or experience, and one's needs can also be satisfied by specific perceptions (when one *sees* what one wanted to *know*). This opens up a possibility that I am going to explore especially in Chap. 5—the possibility of placing human erotetic efforts within a much broader family of explorative behaviors also enacted by other animals, thereby making it possible to learn more about the origin of our erotetic capacity. Moreover, I shall argue below that *the capacity to problematize* cannot be fully explicated in terms of propositions and processing thereof. There is more to our ability to request information, and for that reason speaking of desiderata and their saturata is better suited to the task we face.

That said, someone might argue that speaking of *conditions*, such as C_1 or C_2 above, already renders *desiderata* propositional. In reply, we have to be aware of limitations of any theory of questions. Erotetics is a propositional structure, and this chapter is a propositional structure; therefore we need a way to refer to one's epistemic needs just like, say, ornithologists referring to *foraging*, *scavenging*, and various other patterns of behavior of swallows. The latter, however, does not imply that swallows themselves have these categories in their minds, and so is the case when it comes to C_1 or C_2: they stand for descriptions of possible situations, which *we* need here as parts of our theoretical account of a particular questioner's epistemic needs, but this does not mean that C_1 and C_2 are the exact contents of that questioner's thoughts.

Finally, as already announced, the "is-it-true" schema is special for at least one manifest reason: we do not face two sets of alternative scenarios—one involving the desideratum and recognizing the problematized condition against the background of other possible scenarios, and another involving possible saturata. Instead, we have just one set of alternatives because these questions, in principle, have a different function or a different goal.

The "is-it-true" schema is specific, I argue, since it does not serve to contrast a *predetermined* actual scenario with a variety of possible but unrealized scenarios (possible worlds). Instead, it serves to nail down *which* possible scenario is realized, thus which of the possible worlds is *the* actual world. Here is the relevant desideratum:

(19) *General "is-it-true"* desideratum: Bring it about that I know that C_1 is the case as contrasted with $C_2 \ldots C_n$ *or* that I know that C_2 is the case as contrasted with C_1, $C_3 \ldots C_n$, *or* …, *or* that I know that C_n is the case as contrasted with $C_1 \ldots C_{n-1}$

The saturata are in this case already included *in* the desideratum.

3.7 The "Logic" Behind the Emergence of Questions

I shall now explain what all the things described in the previous section have to do with the puzzle of intuition. Let me start with the following query: how do we actually manage to produce desiderata and then assort appropriate saturata? This issue is not as unproblematic as it may seem at first. For there is no ready-made recipe for contrasting a given condition with alternative scenarios; thus no ready-made recipe for all instances of problematization.

One clear option is that we are *instructed* to make such and such assortments of questions and answers in specific situations, and each such instruction picks some pairs from a huge pool of possibilities. For example, there are two trains, one from Paris to Marseilles and one from Marseilles to Paris, such that the first one is two times faster than the second. A secondary school student is told *what to ask* about. A priori the student could ask about many different things but the instruction is to ask and then calculate at which point the two trains will pass one another. So the student is simply told what kind of knowledge is desired. College students may similarly be asked to problematize the ingredients of a mixture they made as part of their chemistry class, or they may be asked to problematize the impact of institutions on economic growth in developing countries, and so on.

Aside from instructions, in some situations it makes sense to say that questions are *implied* by what one already knows, as well as by other questions (Wiśniewski, 1994, 1995). This point is important insofar as we agree with Hintikka that questions and answers must not be seen as cut off from the broader epistemic context in which they emerge. On the contrary, each answer becomes the base of another question, and each question–answers pair is a move in a wider information-seeking game. The word "game," as Hintikka intends it, is not a mere *façon de parler*. It stands here for a technical posit of game theory. Indeed, according to Hintikka, the pursuit of questions and answers can be represented as a special game. If this characterization is correct, utility is not attributed to one move, but rather to entire strategies. This means that any assessment of the outcomes of one's erotetic pursuits must focus on larger units, not single questions. This observation makes room for question–answers pairs that may seem unintelligible or even somehow incorrect when cut off from their erotetic "surroundings," but which play a role as parts of a larger system (the whole erotetic game). This remark will prove especially critical when we move on to consider philosophical questions.

Therefore, we can speak of different strategies one can adopt in the information-seeking game of questions and answers. Some moves in the game are clearly determined by explicit instructions, but relatively uncontroversial moves may also stem from erotetic "implication." For, given the fact that one always draws on previous knowledge and thus also on previously asked (and answered) questions, there may be cases in which a new question seems to have been produced solely by the machinery of erotetic logic.

Anna Brożek (2011) provides an appealing example.[2] Suppose someone asks Q1: "Who wrote the soundtrack to F. F. Coppola's 'Bram Stoker's Dracula'?" The assumptions, among many, include that there is such a thing as the abovementioned movie, that there is a soundtrack that has been written for this movie, that someone wrote the soundtrack, and so on. Suppose that the answer given by someone is this: A1: "The soundtrack was written by the composer who also wrote 'Orawa,' 'Krzesany' and 'Missa Pro Pace'." Then, according to Brożek, what logically follows from Q1 and A1 is Q2: "Who wrote 'Orawa,' 'Krzesany' and 'Missa Pro Pace'?" The final answer which concludes this game, as Hintikka might say, is A2: "Wojciech Kilar wrote all these pieces of music" or simply "Wojciech Kilar."

It is clear that A1 has the intention of puzzling the questioner, so even if it counts as an answer we are strongly tempted to think that a further question *must* follow. However, does A1 really imply any further questions, and particularly Q2? In principle, each answer can be a platform for further questions, but it can also *not be* a platform of this kind. Think of a situation in which A1 is decoupled from Q1, for example, it is something said to a stranger on the street. Most likely the stranger's reaction will be something like "Who cares?" or, in a more articulate manner, "OK. And what am I supposed to do with this information?" For it is not necessary that A1 itself compels the person to ask any questions. After all, it is only a description of a fact—in this case the fact that the person who wrote the soundtrack to Coppola's "Dracula" also composed several other great works. But why is this fact supposed to provoke any question? An obvious answer would come from Hintikka: you cannot cut off A1 from Q1—all questions and answers play a role as elements of a discourse or epistemic game. On the contrary, when we consider a reasoning like "If Socrates is a human being and all human beings are animals, then Socrates in an animal," the relevant implication would be valid even if it were the only sentence left on Earth. It does not need to be part of any game. Erotetic implication is thus significantly different. However, how did Q1 emerge? Is it based on some previously given answers; when and how did this particular erotetic game begin?

Therefore, aside from the explicit quasi-deductive relations between questions, answers and further questions, there must also be a procedure for generating what might be referred to as *the first question* in a given game. Suppose that a person named Sam comes across a house and asks "Who lives in this house?" and the answer is "John." The mechanism generating this question, outlined by Hiż (1978), has to do with variables and the procedure of substituting either predicates or proper names with variables. Here is a grossly simplified example: we begin from "John lives in the red house," which is the target desideratum. Yet, the questioner, Sam, does not *know* that it is John who lives in the red house, so all that can be said from Sam's standpoint is that "someone lives in the red house" which means "there is *x* such that *x* lives in the red house." Logically speaking, there *is* a description of the fact that *John lives in the red house,* although Sam does know it (let's say that the description exists in the ideal "third" world of Popper's three worlds), and replacing "John" with the variable makes room for a description of Sam's actual knowledge. We can say therefore that "Sam knows that there is *x* such that *x* lives in the red house." And this latter expression is what Sam actually starts from. Now, given that "Sam does not know that it is John who satisfies (is the only person who satisfies) the condition '*x* who lives in the red house,'" we arrive at the relevant desideratum: "bring Sam to know who satisfies (who is the only person satisfying) the condition '*x* living in the red house.' "

The questions for us now is whether or not these logical procedures of substitution and erotetic implication are sufficient to account for all instances of problematization.

3.8 How Do We Know What We Want to Know?

Cautiously put, there is no reason to doubt that *some* questions are, indeed, generated by means of the two procedures outlined above. The procedure of substitution is somewhat similar to what Ingarden (1964) called "variabilization." And once such a question is spelled out, it can give rise to a new question–answer game, leading to various discoveries, accumulation of new knowledge, and so on. Yet the crucial question is this: what drives the said "variabilization"? Why is *this* expression substituted with a variable, and not *that* one, and why is any expression replaced by a variable in the first place? Where does *the need to do so* come from? Questions like these may seem obscure at first glimpse, but let us take a closer look at what they are trying to bring to our attention. Namely, there is no pre-established rule, so to speak, that decides what is supposed to be problematized, and, even more fundamentally, no rule deciding that anything is supposed to be problematized at all.

Therefore, it seems that the pursuit of problematization is something like a spontaneous act of the intellect, perhaps in the sense introduced in Kant's *Critique of Pure Reason*. To see that, first note that it is possible to imagine an alternative procedure even in such a simple case as the red house example cited above, not to mention more complex ones. For when faced with exactly the same red house someone might problematize not the issue of whether it is John or someone else who lives there, but the issue of, say, whether the house is currently occupied by its owner or a renter. Hintikka might argue that it is determined by what S knows, therefore also by what S does not know, and what S seeks to know. Fair enough, but S may know neither that it is John who lives there nor that John is the owner of the house. Both things might therefore be problematized. Actually, the question "Who lives in the red house?" is ambiguous in this respect, for "who" may refer to a concrete person, but it may also refer to a type of a person, for example, to an owner as contrasted with a renter. It is easy to imagine a scenario in which the latter is being asked about, but not the former. Suppose you are taking a walk in a wealthy neighborhood in some alpine village and you wonder whether the properties you see are privately owned or company owned for commercial uses—which is equivalent, more or less, to the owner vs. renter question. All that being said, the puzzle we face is *what drives the pursuit of problematization*.

There is an even more profound philosophical issue that needs to be flagged at this point (one I shall come back to later on)—note that if the very faculty of questioning involves the procedures described above, it in principle hinges on a particular *description* of the facts to be problematized. After all, "someone lives in the red house" is just one description of what one sees, alongside descriptions like "the house is red," "the garden surrounding the house is designed in the Italian fashion," and so on, which shows that different aspects of one and the same perceived scene can be emphasized, provoking different questions. And again, there is no predetermined instruction as to how a given scene should be described, thus brought under concepts, understood and in turn questioned.

A good illustration of this indeterminacy comes from Hintikka and Hintikka (1989) and refers to a fragment from Arthur Conan Doyle's "Silver Blaze." Sherlock Holmes is investigating the case of a dead stable-master and a stolen racing horse, and suddenly he shifts his attention to the guarding dog. The people around him are surprised and he is informed that the dog did not bark in the night-time, which is clearly intended to re-direct the detective's attention to what *all the others regard as important.* The dog's barking, or rather non-barking, seems irrelevant to them. Meanwhile Holmes replies that it is precisely the dog's barking—or, more accurately, lack of barking—that is curious. Clearly, this limits the range of suspects to those whom the dog was familiar with, which is a huge step forward in the investigation. The point is that Holmes problematized an issue that nobody was even aware of. He likely had some sort of heuristics, a pattern of thinking, that he applied to each case he investigated, yet this pattern, if any, can hardly be formalized or algorithmized. Someone might argue that this must be something like "Search for events that normally happen but for some reason did not happen this time," but such a procedure can hardly be effective. Each case is unique, unless our representation of it is very low resolution. Once we switch to a higher resolution, which is likely a good strategy for an investigator, there will be an indefinite number of things that are more or less "normal." How is it possible to pick only those that are relevant? What is more, note that what Holmes did can be described as turning the non-existence of a certain fact, the lack of a certain fact, into a genuine fact. The issue of negative facts merits separate ontological discussion, but insofar as we don't want our ontology to burst at the seams, when *x* does not happen, it is not a fact of *x* non-happening, but a lack of the fact *x* happening. Otherwise, even if there is a definite number of things that do happen over the period of one minute, say in my room, there is certainly a completely indefinite, uncountable number of things that do not happen, and it would be obscure to take all of these non-happenings as facts. Yet, some things that do not happen are *important to us* and once we elicit them, they *as if* become facts—negative ones. So Holmes picked the lack of a specific fact and by problematizing it came up with a crucial negative fact, significant enough to be dwelled upon. The point is, from our perspective, that even if we could think of a problematization algorithm or heuristic referring to facts, there could hardly be an *effective* algorithm or heuristic transforming relevant non-facts into negative facts—for the number of non-facts is undetermined.

The fundamental question that emerges, however peculiar it may sound to an unprepared ear, is how is it possible that *anything* is questioned; that anything is problematized; that anything calls for an answer? Something must be *recognized as a problem* in order to serve as a steppingstone for a question, and the game-theoretic model of discourse gives no clues as to how problematization *as such* is possible. It fails to explain why anybody wants to play the game. Of course, there must be some *interest* involved. The mentioned spontaneity attributable to the pursuit of problematization does not mean that problems just show up out of the blue. As living creatures we problematize conditions that are somehow relevant. However, the real question is, what kind of an interest it is, what is the mechanism of such interest-driven problematization, and so on. All these things shall be tackled later on in the book.

Once something gets problematized, the game starts; but there is no game prior to problematization, and the latter therefore remains a mystery. Certainly, insofar as our epistemic pursuits are located within clearly delineated conceptual frameworks (in Ajdukiewicz's, 1934 or Carnap's, 1950 sense), there is a logic of questioning; a logic generating questions within the said framework. My point is this, however: this logic does not cover our entire erotetic capacity, especially when the proper conceptual framework is not yet decided. In other words, there is something more fundamental, I argue, in our capacity to problematize. The latter point obviously applies to instructions, too. It is impossible to account for *all* assortments of questions and answers, all *desideratum–saturata* pairs, in terms of instructions. At times there are *new* situations, such that there is no ready-made template for them, and there may also be prospects for *novel* recognitions of familiar cases.

A novel recognition, that is, a fresh look at familiar facts, is of enormous importance when it comes to scientific inquiry and intellectual pursuits in general, including philosophy. We can hardly think of a more familiar situation than stepping into a bath and observing the level of water rising. But it was Archimedes who—according to a somewhat mythologized story—looked at this familiar sight in a different way and came up with a novel question–answers pair that–as the legend says–led him to the discovery of what is now known as Archimedes' principle. At the end of the day, there is no more familiar "fact" than the very *presence* of the world around us, but it was Parmenides and others who coined the notion of *being* which enables us—however peculiar it sounds from the logical standpoint (see the Chap. 4)—to problematize this most familiar "fact" of the world's *being there* and our being-in-the-world (see Chap. 5).

Going back to our previous example, perhaps the tragic case of Kennedy's assassination is no longer appealing at this point, since killing a sitting US president is so unusual that it can hardly be ignored, meaning not recognized as *a problem*. Note, however, that even if it strikes us as obvious that the assassination of president Kennedy is a turn of events that calls for a special investigation, it strikes us this way *because* our spontaneous reaction to what happened in Dallas that day recognizes it against the background of a more usual scenario (in which the president is not killed). This spontaneity poses a real challenge for any theory of questioning. Moreover, although the need to problematize Kennedy's assassination seems obvious, there is in fact a large variety of things that such problematization may target. I can naturally recognize the course of action in which Kennedy gets killed in contrast to a variety of different courses of action in which he does *not* get killed. However, I can also see it in contrast with a variety of courses of action in which he is more or less heavily wounded, a course of action (in fact, also a variety) in which he is, say, stabbed, or killed by a car bomb, as well as a variety of scenarios in which someone else gets killed, and so on. All of these problematizations may result in questions, and some of them have to be articulated as part of an investigation: for example, why someone wanted to kill Kennedy, and not, say, the Texas governor John Connally; is it *really* the case that the shooter wanted to kill Kennedy, not Connally (perhaps Connally was the target?), and so on.

Now, what differentiates a good investigator from a bad one in this context? A tempting answer, harking back to the Holmes example, is this: a good investigator sees more issues that are relevant to the investigation than others do. He is therefore able to plausibly problematize issues that are relevant, yet unnoticeable to less gifted people. Yet the good investigator cannot problematize everything one might think of, since it would obviously be pointless. What then is this peculiar gift that the good investigator has? Generally speaking, the puzzle is *how we come up with our desiderata*; how we come to problematize the relevant issues; how we combine questions and answers in the single engine powering our cognitive endeavors. To be clear, I am not asking about the actual cognitive *mechanism* responsible for the procedure, which can be tracked down empirically. I am asking a conceptual question that precedes the empirical one, namely: *What* is the intellectual faculty, the *faculty of problematizing*, in itself?

Given all that has been said so far in relation to erotetic logic and propositional procedures that can in many (but not all) cases effectively generate questions, I would like to now try out a different route. I will begin with an observation based on everyday phenomenology, then formulate a posit whose sole purpose it to give a name to the observed phenomenon while at the same time dispelling some of our initial ignorance as to how the said posit works. In the final chapter we shall try another approach—one that is new to erotetic considerations but quite well established in other domains—which will enable us to at least partly fill the gap in our understanding.

So, aside from the erotetic procedures outlined earlier, another option is that some perceptions or experiences *afford* certain problems to the questioner, just like, according to the famous proposal by James J. Gibson (1979), they afford possible courses of behavior by "offering" themselves as usable things or tools. In other words, problems show up or manifest themselves to us, relative to our needs and interests. However, the receiver, determined by specific needs and interests, must also be appropriately equipped to understand the "offer." Hence, if the world provides food for thought, one needs an intellectual faculty to "digest" it. In the case of human beings, this particular faculty is armed with sophisticated linguistic and computational tools, the already mentioned logic of questions, but I argue that it must not be identified with these tools. At the bottom of this toolbox is a *residuum* that I refer to as *erotetic intuition*. It is not some mysterious extramundane gift from the gods, but a product of evolution, most likely one of the pivotal factors in the early evolution of hominids. Although it is no longer accessible in its "pure," pre-linguistic form, nonetheless it deserves to be named and given a positive characterization in its own right (noteworthily, in a similar spirit Chomsky, 1987 speaks of what he refers to as the *science-forming faculty*, but the latter would already be a very sophisticated instantiation of the faculty I propose to talk about).

This chapter provides a negative characterization. I argue that the postulated residual erotetic capacity underlying every act of problematization cannot be explained in propositional terms. The term "intuition" must also be approached as part of this via negativa, as it were. At this point the only positive thing we can say about the targeted capacity is that it makes problems or questions manifest; and since it cannot boil down to propositional processing—as I show below—it must be something else. "Intuition" is a traditional, rough-and-ready term that seems to spell it out quite well.

3.9 Intuitions: Percepts or Beliefs?

Beginning perhaps as early as Plato's "Seventh Letter," up to Wittgenstein's "Tractatus logico-philosophicus" and "Lecture on Ethics," philosophers have recognized that there is sometimes more to our cognitive capacities than we are able to describe. It may be the case, they say, that some contents simply *manifest* themselves to us, full stop (see Wittgenstein, 1922). Indeed, the term "intuition" has been used in the philosophical literature to cover our ignorance as to how this manifestation comes about. It is said, for example, that if a belief is based on sensory awareness of some entity or fact, then it is *not* intuitive. If a belief is supported by a mathematical proof, it is *not* intuitive either. We thus arrive at the idea that intuition is a judgment that is not based on any clearly defined process of reasoning (Gopnik & Schwitzgebel, 1998) or perceiving. This belief may stem from an understanding of specific propositional contents (Sosa, 2007); for example, I intuitively know that Mary is a widow if her husband died. Alternately, in the Cartesian spirit, intuition might be a specific content that intellectually manifests itself *clearly and distinctly*. For example, I intuitively know that adding two positive number produces a positive number. Pust (2019) concludes that a single intuition is either a belief or a *sui generis* mental state in which some content presents itself to the intellect; respectively, it intellectually seems to the subject that something is the case (for an overview, see Booth & Rowbottom, 2014).

Some scholars, for example, Elijah Chudnoff (2011), seem to take the term "intuition" as the name of a proper cognitive faculty. Driven by this presumption, Chudnoff aims to provide a positive characterization of intuition by drawing on the analogy between sensory awareness of concrete entities on the one hand and intellectual awareness of abstract entities on the other (for more on philosophical intuitions see, e.g., Deutsch, 2010; Cappelen, 2012; Nado, 2014).

To begin with, Chudnoff distinguishes:

Doxastic views: intuitions are either judgments or inclinations to judge.

Perceptualist views: intuitions are pre-doxastic experiences representing "abstract matter as being a certain way" (Chudnoff, 2011, p. 626). They are similar to perceptual experiences, yet instead of representing concrete spatiotemporal entities, like the ones we see or hear, they represent abstract entities or contents.

Chudnoff's own view belongs to the perceptualist camp. He defines intuitive knowledge in the following manner:

> Intuitive Knowledge (IK): If an intuition makes a belief that *p* based on it amount to knowledge, it does so in virtue of (1) being an experience in which it intuitively appears to you that *p*, and (2) being an experience in which you are intellectually aware of an item *o*, such that (3) *o* makes *p* true. (Chudnoff, 2013, p. 375)

Chudnoff argues against doxastic views in a way that parallels the arguments against the propositional content of perception. I shall also adopt this strategy below, yet although I lean more toward the perceptualist approach, for reasons that shall become clear shortly, unlike Chudnoff I don't consider the issue in question that fundamental. For it seems that the *doxastic vs. perceptualist* distinction refers not so much to intuition (if any) itself, seen as a certain cognitive faculty, as to its various *manifestations*.

Let me substantiate the latter point. Consider the doxastic approach first. Suppose we concede that a particular judgment is the *product* of a certain cognitive capacity or faculty. However, the very notions of *inclination*, or, to put it in a somewhat Kantian way, *the power* of judgment, are in fact theoretical postulates. The latter means that we never encounter an inclination to judge or a power/capacity/faculty of judgment *as such*; we are only ever dealing with *a* judgment, expressed in the form of a proposition, which we interpret as a manifestation of a particular inclination or capacity.

As far as experience is concerned, we face an analogous scenario: for we can think of experience as something concrete available to us immediately, that is the product of a presumed sense or sensory capacity (the capacity of experiencing), of which we learn only when faced with its products: with the concrete experiences that fill our everyday lives. Thus, we assume that there is a *sense* or sensory capacity which produces actual perceptual experience. The capacity in question is a certain theoretical posit. Crucially, speaking of *sense* is not equivalent to speaking of the relevant sense organ. The organ is a part of the body, composed of proteins, among other things, capable of receiving and processing a specific kind of energy, be it the energy carried by electromagnetic waves or mechanical waves, or the energy of chemical reactions. Speaking of the five *senses*, however, refers to knowledge or content generators which are somehow (exactly how has been a mystery since the time of Descartes) based on the physico-chemical machinery just mentioned, but not identical to it (at least when conceptually analyzed). So, the senses or sensory capacities are theoretical postulates which are supposed to answer the question of how and why we experience the world the way we do, how these experiences contribute to our knowledge of the outside world, and how it is possible that have an experience in the first place.

What I mean is that when intuition is considered in the first place, it should be *provisionally* targeted as a faculty responsible for generating specific content, or, in short, as a specific generator of content, manifesting itself in the guise of experiences *and* judgments, possibly among other things. One can be a pluralist with respect to manifestations and focus one's attention on what constitutes the genuine puzzle in this context, namely the intuitive capacity itself.

3.10 The Erotetic Capacity Is Not Propositional Machinery: The Assumptions Underlying the Main Argument

I will now roll out an argument in favor of the announced negative claim that we cannot fully account for our erotetic capacity in terms of propositional machinery within a fixed language. In short, *erotetic capacity cannot boil down to* erotetic logic.

My argument is based on two assumptions.

Assumption 1 says, following Tim Crane (2006), among others, that there is a difference between the (phenomenal) content of experience (perception) itself, that is the state of being directly acquainted with entities in the world, and the propositional content of a *description* thereof. This view goes against those who argue that perceptual content is in itself propositional, for example, Susanna Siegel (2006, 2010, 2016), who appealingly proposes that the content of perception is not like the content of a bucket (thus full of sense-data or impressions) but rather like the content of a newspaper. Hence, when I see a scene, I see, for example, *that* the tree is green; *that* the bird is sitting in the tree, and so on. The identification of perceptual or experiential content with propositional content is articulated most explicitly by Alex Byrne who says that "perceiving is very much like a traditional propositional attitude, such as believing or intending," to the effect that "when one has a perceptual experience, one bears the perception relation to a certain proposition *p*" (Byrne, 2005, p. 232). The assumption might seem quite controversial, given that there is a huge debate over perceptual content, and for that reason I shall provide some arguments for it below. Besides, it may not be clear upfront how this assumption is related to the puzzle of erotetic intuition, but this is made clear in the proper argument rolled out below in 3.13.

Assumption 2 says that the meaning of a proposition belonging to, and therefore also the meaning of a question generated within, a given description of experience depends on the whole structure to which it belongs; once one modifies the description, for example, by adding new propositions to it, the meaning of the description as a whole, and therefore the meaning of questions internal to the description, changes. In step with the standard view, the meaning of a proposition is its truth-conditions. On the other hand, desiderata can be seen as the meanings of questions. Therefore, putting the assumption in a different way, it says that the truth-conditions of a proposition within a description of experience are dependent on the truth-conditions of all the other propositions belonging to the description. Once the latter are altered, the former change, too. And once the meanings within the description as a whole change, the desiderata existing within it change as well.

3.11 Arguments for Assumption 1

I shall now defend both of these assumptions, beginning with the first one. The notion of experience or *content of* experience is assumed as given, stipulated ostensively: the content of experience is what you see around you, what you feel about what you see, and so on.

Argument 1. From the impossibility of standing in logical relations. Firstly, as Crane (2006) points out, if perceptual contents were propositional, they would stand in logical relations. We would have been able to, for example, negate them or disjoin them as we do with propositions. Crucially, note that logical relations are productive in the sense that if I have *p* and add negation, thereby obtaining *non-p*, I produce a new proposition. If I have *p* and I have *q*, and then come up with *p and q*, I have a genuinely new proposition. However, just as with the content of pictures, according to Crane, the content of experience itself cannot stand properly in such relations, where "properly" means, in my view, "productively" (among other possible things). For example, if certain phenomena are presented to me as I look through the window in my living room, I cannot negate them and get new phenomena. Admittedly, I can always say "it is not true that I perceive a tree," but what I negate is a *description* of my experience, not the experience itself. As result of the negation, crucially, I obtain a new description, but nothing changes in my experience. I am unable to produce new experiences by means of logical operations.

Argument 2. From the immunity of experiential content. The last point from Argument 1 leads to the argument invoked by Chudnoff (2013), frequently quoted in the literature, namely that perceivers cannot help perceiving certain things even if at the propositional level they know that a given experience is not accurate. Actually, it's a *demonstration*, not a proper argument. For example, we see the two arrows in the Mueller-Lyer graph as different in length despite the fact that we know that they are equal. So if experience exists in the domain of propositional contents together with descriptions thereof, the immunity of the former to any modifications made on the latter is a serious problem. Actually, it suggests that the content of experience is different in its nature from propositional content. This in turn raises another issue, namely the difference between accuracy and truth.

Argument 3. From the difference between accuracy and truth. Experience can be inaccurate with respect to real things. Defenders of the propositional-content account of perception (see Siegel, 2006, 2010) use this fact to justify the claim that the content of experience can be deemed true or false, and must therefore be propositional. However, although the idea of the inaccuracy of experience is perfectly clear, it is by no means obvious what the opposite condition, thus accuracy, means. If I say that, for example, $4 < 5$, it is perfectly clear what it means for this proposition to be true. And it cannot be more or less true. It is true, full stop. Propositions referring to everyday issues may not be so precise, but in most cases we have a fairly clear grasp of what it means when they are true, for example, when someone says "the bank is closed today." Compare it to experience. What would "accurate experience" stand for? The walls in my living room are white but when exactly do I see them accurately taking into account the changes in lighting? All I can do is to conventionally define some standard and use it as my reference point in assessing accuracy. Or take, for instance, shape: what could it mean that I experience the shape of, say, the desk I am sitting at accurately? As I look at it from a certain angle, for example, its top does not look rectangular to me. But *even if* I formally determine a point in space from which the top would present itself with perfect accuracy, I must realize that *I* am not a point but a bulky real thing with two eyes located in slightly different places, the data produced by which must be further processed to come up with a unified representation, and therefore, in fact, I can never occupy an ideal *point* of view. So, firstly, we can never experience things with perfect accuracy. Secondly, and crucially, unlike truth (as usually conceived) accuracy is a matter of degree (see Crane, 2006). Therefore, even if it makes sense to say that experiences can be inaccurate, nothing can be automatically or unequivocally concluded about the alleged propositional character of the content of experience.

Argument 4. From the relative richness of experiential content. Descriptions are selective and, in this sense, more or less coarse-grained. This applies to experience, too, but there seems to be a qualitative difference between the two capacities. You can pick some fragments or aspects of experience and "give" them, as Crane says, in the guise of a proposition. But there is always something left, and incorporating a piece of new experiential content may change the entire description. Think of a painting such as Caravaggio's "Calling of St. Matthew." There is a controversy as to whom Jesus is actually calling, so one description says "Jesus is calling the young man," while another says "Jesus is calling the old bearded man." There is also an enormous richness of experiential content in the painting, so that a description can emphasize the incredible beam of light: but is it the finger that does the calling, so to speak, or is it the light which manifests God's will? Interpreters debate this, too.

The possibility of changing a description of experiential content is crucial for my main argument in favor of erotetic intuition which I will roll out shortly. Before I get there, however, I need to make one crucial remark. Someone might argue that if experiential content is not propositional while description is, the two can hardly be *compared*. In other words, we can have the vague impression that we compare them when we say that the allegedly non-propositional content of experience is always richer than the description, but if we had to make this comparison precise we would have to find a platform shared by both experiential and propositional contents, and a tool with some metrics to measure the content of each. This, however, is hardly a thinkable scenario—for what could this shared platform be if the two types of content are supposed to differ so fundamentally? Therefore, the argument would conclude, there is no substance in the richness argument (insofar as the asymmetric relation "to be richer than" implies the symmetric relation "to be comparable to") and thus no reason to argue against the propositional character of experiential content based on it.

However, this is why I have called this argument the argument from the *relative* richness of experiential content. It could also be dubbed the "erotetic argument," much in the spirit of this whole investigation. So the argument does *not* postulate that experience is *actually more abundant*; rather it means that one can always "ask" experience to provide more information. Suppose you are not satisfied by the answers provided by your description, be it, for example, a description of Caravaggio's masterpiece. In other words, within the conceptual framework you decided to use to describe the picture you formulated certain desiderata and provided possible answers to them, yet if your desiderata are still not fully satisfied (you are not happy with the options you obtained), you can always look to the picture for more options and add new answers, amend your desiderata, or articulate new ones. Take a different case, for example, the one invoked by Schaffer (2007). If the desideratum is "bring it about that I know that there is a canary in the tree or I know that there is a goldfinch in the tree," and you got stuck not knowing how to decide, you can first learn more about the two species and then "ask" your experience for more information by articulating a new, auxiliary desideratum such as "bring it about that I know this bird has a black spot on its head or I know that the bird has no spot on its head." You will most likely get the needed answer. On the other hand, if you were only dealing with a description of the bird, that would be a ready-made construct such that you could not get more than its author put there. If the author did not say anything about the black spot, you could not get any information about it (for the sake of the argument, let me limit my attention to *explicit* descriptions in clear terms, leaving aside more profound texts that can be *interpreted* in various ways).

The point of this argument therefore is *not* that there is pre-given content in experience which is more abundant than the content of a description. The point is that *being guided by descriptions*, by questions and answers, you can always *elicit or enact more content* in experience if you need more information. Metaphorically speaking, even if some experiential contents have to be called out by a description, there is virtually no limit to how experience can respond to that calling. Aside from obvious limitations imposed by our cognitive apparatuses, their physico-chemical and biological functioning, experience is unlikely to "say" something like "that's all I have." Meanwhile, a particular description always provides a fixed portion of information. Therefore—my argument concludes—it is impossible to equate experiential content with propositional content. The latter's function is to bring the former into intelligible presentation, as we might say paraphrasing Husserl, yet the former can hardly be fully exploited by any number of propositions.

Someone might argue that the idea of the relative richness of experience is epistemologically risky since it may encourage attempts to justify any belief one can come up with. In other words, the worry is that experience would "allow" for too many beliefs, including those lacking any real empirical support. Siegel gives a nice example of the distinction between the proper propositional content of perception and more derivative beliefs referring to perception, a distinction that seems crucial from the perspective of perception-based *knowledge*:

> there are the contents that a perceiver comes to believe on the basis of her perception, on the one hand; and there are the contents properly attributed to perception itself, on the other. Consider the content:
>
> (#) that Ms Elfenbein went to Ankara.
>
> Suppose that you believed that Ms Elfenbein is out of town only if she went to Ankara. And suppose you went to her house, and found that her curtains were drawn, that her mailbox was overstuffed, and that repeated ringings of her doorbell went unanswered. Then you would reasonably come to believe (#) on the basis of your visual experience. But (#) does not seem to be properly attributable to your visual experience itself (in this case). We seem to be able to distinguish what is presented perceptually from what we go on to believe. We can draw this distinction, no matter how much or how little overlap there may be between them. (Siegel, 2006, p. 481)

So the worry may be that if we abandon the notion of propositional content of perception and distinguish pre-propositional content of experience itself, the line stressed by Siegel cannot be drawn.

I disagree. The line can still be drawn, but it will be more contextual. If I describe what I see as "the mailbox is overstuffed," this seems to be a description of *exactly* what I see, whereas something along the lines of "Ms Elfeinbein left on vacation" is more like an interpretation of what I see. Yet what belongs to the categories "exactly what I see" and "interpretation of what I see" depends on how much I want my experience to "tell" me. Namely, if descriptions can always "ask" the experiential content for more information, as argued in Argument 4, whether experience tells me something only about the mailbox or about Ms Elfenbein's absence depends on what I "ask" my experience to "tell" me. Admittedly, within any description of experience, respectively within any set of beliefs based on experience, we can contextually distinguish a group of propositions describing the purported simple facts of the perceived scene, for example, "the mailbox is overstuffed" or "the curtains are down." They are somewhat like the protocol sentences discussed vividly in early analytic philosophy and methodology of science, and could be taken as constituting the *basic matrix* of a given description of experience. These protocol-like propositions then stand in logical relations with one another and with other beliefs, for example, "the mailbox is overstuffed and the curtains are down" or "if the mailbox is overloaded then Ms Elfenbein is not at home." Hence, by endorsing the non-propositional content of experience and different descriptions thereof we are not automatically committed to an extravagant belief that anything goes when it comes to descriptions and that there is no sense in believing that one description is more adequate than another. Referring to the basic matrix is supposed to stabilize our descriptions of experience and anchor them in commonly accepted facts. Crucially, however, the point is that there is *no universal recipe* for distinguishing these protocol-like propositions from more complex descriptions loaded with expectations, predetermined beliefs, and so on. That being said, this contextual distinction is still useful in each *specific* case of description of experience (I shall go back to Siegel's example below).

3.12 The Argument for Assumption 2

Based on the distinction just made, Assumption 2 can be re-articulated. It says that in the case of any description of experience, call it D1, if a new protocol-like proposition is added to the matrix of D1, then at least *some* other propositions belonging to D1 change their meanings, that is, their truth-conditions, since their truth-conditions depend in part on their relations with the truth-conditions of the newly added propositions. This implies that the meanings of *all* the questions produced on the basis of these modified propositions change, too. D1 therefore becomes a different description, be it D2. In other words, if a new perceptual fact is elicited by a protocol-like sentence (just to be clear, this is a sentence conventionally agreed upon to serve as part of the matrix of a description), we end up with—in principle—a new description. The phrase "in principle" indicates that the change at stake does not have to be of huge practical significance (it may not be noticed), but this is a different issue. I also have to stress that this assumption does not equal full-blown semantic holism as famously proposed by W. v. O. Quine or Donald Davidson; here my concentration is solely on descriptions *of experience*.

Here is an argument for Assumption 2; in fact, it is, traditionally speaking, a demonstration. To begin, think again of Siegel's scenario of visiting Ms. Elfenbein. Suppose I approach her house with a plan to speak to her, and I see a magazine in a plastic bag hanging on her door-handle, left by the mailman. I form a belief, a protocol-like proposition, based on it, that there is a magazine left at Ms. Elfenbein's front door. It becomes part of the matrix. Based on that and my previous knowledge of how the world used to work, I form a belief that Ms. Elfenbein is not at home. This belief has many possible truth-makers: if she is Ankara, the belief is true but if she just went to the closest grocery store, the belief is also true. Now, suppose I take a closer look and add new protocol-like reports to the matrix, such as "Ms. Elfenbein's mailbox is overstuffed" and "the magazine is *Philosophical Quarterly*," which set up spontaneous bonds with some newly extracted parts of my previous knowledge of Ms. Elfenbein, such as her impeccable sense of order (which wouldn't allow her to ignore the mess under her mailbox) and my knowledge of *Philosophical Quarterly*'s being, as the name indicates, a quarterly (thus a periodical mailed out to subscribers only four times a year, which makes it plausible that it was not hanged on the door-handle just this morning but perhaps much earlier). Based on these assumptions I can still believe *that Ms. Elfenbein is not at home*, but the meaning of this seemingly identical belief is now different, which may be represented by the fact that the scenario of her going to the closest grocery store is *not* considered a potential truth-maker any more. So adding new protocol-like propositions to the basic matrix has changed the truth-conditions of some of the propositions previously included in the said description, thereby changing their meaning. In addition, if we think of a desideratum such as "bring it about that I know where Ms. Elfenbein is," its actual sense was different before I noticed the mailbox and the *Philosophical Quarterly* hanging on the door-handle, for it previously referred to a huge list of alternatives such as "bring it about that I know that Ms. Elfenbein went to Ankara or that I know that Ms. Elfenbein went to Paris or that I know that Ms. Elfenbein went do the grocery store… etc." After having noticed the two additional phenomena, *seemingly the same* desideratum expressed a much more limited number of alternatives, but perhaps also an additional one that by any stretch of the imagination was not on the previous list, such as "Ms. Elfenbein is in hospital."

3.13 Erotetic Intuition in Response to Erotetic Skepticism

Finally, provided that Assumptions 1 and 2 are set, here is my argument for the claim that there must be an intuitive capacity to problematize selected parts of our surroundings *directly*; to "see" problems there.

Assumption 1. There is a difference between the content of experience and the content of a description thereof. *Step 1.* Based on Assumption 1, suppose one wants to improve description D1 driven by dissatisfaction with the questions and answers that can be produced within D1.

Assumption 2. If a new protocol-like proposition is added to the matrix of D1, thereby giving rise to D2, then—in principle—all the propositions previously belonging do D1 change their meanings in D2. This implies that the meanings of *all* the questions produced on the basis of these modified propositions change, too. *Step 2.* Suppose now—as a proponent of the purely logical approach to questions would argue—that all desiderata, thus all questions within D1 and D2 are generated by logical operations on the propositions belonging to D1 and D2.

Step 3. If the transfer from D1 to D2 is to succeed, that is, to increase the amount of information obtained from experience, one has to restore the relevant questions existing in D1 within the new context of D2, that is, re-articulate one's *previous* epistemic needs, within D2.

Step 4. If some of the questions internal to D2 are to be recognized as expressing the same epistemic needs as those in D1, so that D2 can be *expected to do a better job* than D1 (which, after all, was what motivated us to move from D1 to D2), *they must be expressions of the same desiderata.*

Conclusion: if Step 2 is the case, all the desiderata within D2 must be generated by suitable logical operation within the boundaries of D2; full stop. Therefore, Step 4 is impossible.

This is a paradoxical situation. On the one hand, moving from D1 to D2 was propelled by the need for more knowledge than D1 was able to provide, while on the other hand, there is no clear link between the old set of desiderata and the new ones that would actually allow the questioner to get more data *referring to the same epistemic needs.*

Yet, suppose that the questioner does replace D1 with a new description D2. Then the questioner reformulates all the previous questions in this new propositional context, meaning that they are *recognized by the questioner* as articulations of identical epistemic needs. Here is a puzzle that might be brought up by an *erotetic skeptic*, if I can name the character this way: *how* can the questioner *know* that the new questions express *the same epistemic needs*? What are these mysterious residual epistemic needs, as if binding D2 with D1, if all the actual desiderata belonging to D2 must rely, given Step 2, solely on the propositions from D2?

Someone may argue that the worry expressed by the skeptic—let us call it *erotetic skepticism*—is far over the top and purely speculative. In real-life cases, descriptions are not isolated islands. There are social practices linking descriptions with one another, so there is no *justified* worry that by switching from one description of an experience to another, the questioner loses ground completely and cannot be sure that he is still within the same setup of problems and epistemic needs. Given the social practices we share, these worries are merely idle speculation.[3]

In response, let us note an interesting fact about this the down-to-earth reply to erotetic skepticism. Namely, the very idea that the questioner *may not know* whether the desiderata articulated in D2 are the same as the ones in D1, and whether or not D2 as a whole refers to the same experience as D1, seems absurd *only if we assume upfront* that the questioner *does* have access to at least some contents of their experiences independently of this or that description, and—most importantly—that this access channel, so to speak, is of some epistemic significance. In other words, the alleged absurdity of erotetic skepticism strikes us with such power due to the fact that we already assume that our experiences do *afford* something to us, *including the affordances of problems to be solved*, in a sense similar to Gibson's (1979).

The latter means that we have to reject Step 2 in order to get rid of the paradoxical conclusion. Yet there is the question of what this supposed direct and epistemically significant *erotetic link* to the contents of our experiences actually is.

We can only provide, as announced, a provisional answer at this point. Namely, we need to posit a specific cognitive faculty, a kind of *intuition*, that is, the capacity to intellectually "see" problems to be tackled, more or less in the way described by Chudnoff. It is responsible for coming up with desiderata, for problematizing different things or phenomena in our surroundings. This capacity of direct, *as-if-perceptual* problematization of different phenomena in our surroundings makes it possible to make the transfer from D1 to D2 without losing ground; thus to re-articulate in principle the same desiderata within a new description.

We can define the said capacity referring to each type of desiderata discussed earlier; however, it is sufficient for now to do this with respect to the desideratum in its general form. So, I propose that:

(20) Erotetic intuition: If C_1 is a condition realized in the real world, and s is a questioner, s has erotetic intuition about C_1 if s is perceptually afforded with C_1 *as contrasted with* a (fully or partially specified) set of alternative conditions $C_2,\ldots,C_n$. In Chudnoff's terms, s has erotetic intuition about C_1 if s experiences C_1 and it *intellectually appears* to s that there are alternative conditions $C_2,\ldots,C_n$.

3.14 The Puzzle of Directness: Stepping Out of Our Heads

Erotetic intuition has been portrayed as the capacity enabling us to directly "see" things that "call" for an explanation; to *directly see* facts as contrasted with alternative scenarios. Yet, this *directness* is actually quite challenging philosophically. For there is a debate whether or not experience *can* provide us with direct access to the external, mind-independent stuff in the first place. Scarce representationalism in the philosophy of perception, as it was dubbed by Susanna Schellenberg (2011), denies any such direct access, claiming that perceptual content is fully constituted by representational contents. The other scarce position—relationism—makes the case that "the direct objects crucial to understanding perception (…) are perfectly mind-independent physical things" (Brewer, 2007, p. 89), which means that "perceptual experience is not representational, but rather fundamentally a matter of a subject being perceptually related to mind-independent objects, properties, events, or the event in which such relations obtain" (Schellenberg, 2011, pp. 714–715). Now we can see what is at stake in this debate. While relationists emphasize direct access to the external, mind-independent world as the crux of perception, scarce representationalists champion indirect access, meaning that we are in fact dealing with the products of an internal cognitive machinery processing stimuli.

This debate is serious for it is actually about the ultimate grounds of knowledge as such. In other words, it is about the very possibility of our percepts, thoughts, theories being *acquainted with the real world*. The high stakes were nicely unpacked by Imogen Dickie in a different context, that is, with reference to so-called *acquaintance-based thoughts*—those that "we are in positions to have because we are acquainted with the objects they are about." Dickie's argument goes like this: "All other thoughts about particulars depend for their representational content on acquaintance-based thoughts. So there must be acquaintance-based thoughts. For otherwise none of our thoughts would be about anything at all" (Dickie, 2010, p. 213).

Representationalism has a difficult time trying to account for how it is possible for perception to directly link us with the surrounding world and give rise to acquaintance-based thoughts which are fundamental to our knowledge. Meanwhile, relationism has trouble with another aspect of perception, which, on the other hand, is accounted for quite convincingly by representationalism, namely the fact that the world is always perceived a certain way; that we are always dealing with a certain layout of the world that fits our cognitive tools, depends on context, and so on. Postulating a variety of internal representations seems to be a natural response to this, whereas it is not clear how a direct bond with the real thing, postulated by relationism, could also be *a certain way* (it seems that either I grasp the thing or not; there is no place for aspects).

The take for us here is this: *directness* poses a serious philosophical challenge in the philosophy of perception and in epistemology, and the same applies to erotetics. The postulated erotetic capacity to problematize is supposed to be direct, while at the same time also working *a certain way*, for there is always a variety of desiderata we can come up with, bearing the baggage of who we—the questioners—are and what our capacities, needs, goals and tasks are. How are we to account for all these factors within a unified conceptual framework?

There is one common trait in both of the cited views: they are static. Either perception merely answers the presence of external objects, as relationism proposes, or as representationalism sees it at the other extreme, the entire active part of perception, that is, the computational part in this case, is locked up in the head, thereby also deeming the perceiver placed in a specific environment a static recipient of stimuli. This is mirrored by the alternative we face in the domain of erotetics: either the pursuit of problematization boils down to internal logical processes involving propositions, rendering the questioner static in his environment, or—if we follow Chudnoff's intuitionism—problems simply manifest themselves to us. The puzzle is, however, that on the one hand we have argued that problematization can hardly be only about these propositional procedures, yet on the other hand, our intuitive grasp of problems simply waiting to be apprehended is also somewhat mysterious and unlikely. What counts *as a problem* clearly depends, in one way or another, on the person *whom* it calls upon for an answer/solution. What constitutes a problem for a man may not be a problem for, say, a wolf, and what constitutes a problem in the wolf's environment may not be a problem for a fly.

As such, the pursuit of problematization must not be targeted, I presume, as something we merely exercise in our heads. Instead, erotetic intuition must be taken as a faculty attributable to living, acting creatures coupled with their surroundings. To put it simply, our erotetic capacity must be grounded in what we *do in the world* and how we do it.

This is the moment for a decision. A still unexplored option opens up before us, for there is a newly emerging paradigm in philosophical thinking about cognition, perception, and consciousness which makes the case that being directly presented with real things in our surroundings does not come cheap, so to speak. Metaphorically speaking, following Alva Noë (2012), the presence of the world must be earned by us based on how we act in the world. This paradigm comes under many names such as "embodied cognition" or "enactivism," among others.

So, as for the announced decision: in Chap. 5 I shall explore the possibility of an embodied account of the pursuit of problematization. This will shed some light on the gap in our understanding of cognition and knowledge, especially when it comes to the issue of directness.

3.15 Conclusion: What's Next

This chapter set forth the notion of erotetic intuition. This is the capacity for problematization. I argue that while it makes a lot of sense to study erotetic logic, thus the propositional machinery of questioning, following scholars such as Hamblin, Hiż, Rescher, and Hintikka, it is still not enough to account for the phenomenon of problematization. The semi-formal clarity of erotetic logic enables us to articulate the issue in terms of different kinds of questions and desiderata. That said, I argued, referring particularly to the problem of the directness of our erotetic capacity (as compared to the problem of perceptual directness), that we need to take a step forward and re-direct our attention from the sole focus on the internal processing of information, including requests of information (desiderata), toward the pursuits of questioning and problematization as proper parts of cognition thought of more broadly as an evolutionarily shaped, embodied activity, necessarily coupled with an environment.

There is one thing, though, that needs to be addressed before we move in this new direction (in Chap. 5). For, given our initial concerns from Chaps. 1 and 2, someone might ask, how exactly do philosophers exercise their erotetic intuition? This puzzle is going to be the topic of Chap. 4.

Notes

1. Hintikka adds that the *how-* and *why*-questions constitute a special group which does not behave exactly like *wh*-questions, but he only limits his attention to aspects of *how-* and *why*-questions that overlap with those of *wh*-questions.
2. Brożek refers to Wojciech Kilar, as I do, but her questions are different. The reference to Coppola's well-known movie is merely intended to make the example more accessible.
3. Note that the same point could in principle be made with regard to any kind of philosophical skepticism. After all, is Descartes' skepticism about the existence of external reality less speculative? Is Kripke's (or Wittgenstein's; or "Kripkenstein's," as it sometimes referred to) semantic skepticism regarding rule following less radical and less speculative than our erotetic skepticism? These are, of course, rhetorical questions. Skepticism by design is radical and speculative.

References

Ajdukiewicz, K. (1934). Das Weltbild und die Begriffsapparatur. *Erkenntnis, 4*(1), 259–287.

Åqvist, L. (1969). Scattered Topics in Interrogative Logic. In J. Davis, D. J. Hockney, & W. K. Wilson (Eds.), *Philosophical Logic* (pp. 114–121). D. Reidel.

Åqvist, L. (1971). Revised Foundations for Imperative Epistemic and Interrogative Logic. *Theoria, 37*, 33–73.

Belnap, N. (1983). Approaches to the Semantics of Questions in Natural Language. In R. Bauerle, C. Chwarze, & A. von Stechow (Eds.), *Meaning, Use, and Interpretation of Language* (pp. 21–29). de Gruyter.

Belnap, N., & Steel, T. (1976). *The Logic of Questions and Answers*. Yale University Press.
Booth, A. R., & Rowbottom, D. P. (Eds.). (2014). *Intuitions*. Oxford University Press.
Brewer, B. (2007). Perception and Its Objects. *Philosophical Studies, 132*(1), 87–97.
Brożek, A. (2011). *Theory of Questions. Erotetics through the Prism of Its Philosophical Background and Practical Application* (*Poznań Studies in the Philosophy of the Sciences and the Humanities* vol. 99). Rodopi.
Byrne, A. (2005). Perception and Conceptual Content. In E. Sosa & M. Steup (Eds.), *Contemporary Debates in Epistemology* (pp. 231–250). Blackwell.
Cappelen, H. (2012). *Philosophy Without Intuitions*. Oxford University Press.
Carnap, R. (1950). Empiricism, Semantics, and Ontology. *Revue Internationale de Philosophie, 4*, 20–40.
Chomsky, N. (1987). *Language and Problems of Knowledge: The Managua Lectures*. MIT Press.
Chudnoff, E. (2011). What Intuitions Are Like. *Philosophy and Phenomenological Research, LXXXII*(3), 625–654.
Chudnoff, E. (2013). Intuitive knowledge. *Philosophical Studies, 162*, 359–378.
Crane, T. (2006). Is there a Perceptual Relation? In T. S. Gendler & J. Hawthorne (Eds.), *Perceptual Experience* (pp. 126–146). Oxford University Press.
Deutsch, M. (2010). Intuitions, Counter-Examples, and Experimental Philosophy. *Review of Philosophy and Psychology, 1*, 447–460.
Dickie, I. (2010). We Are Acquainted with Ordinary Things. In R. Jeshion (Ed.), *New Essays on Singular Thought*. Oxford University Press.
Gibson, J. J. (1979). *The Ecological Approach to Visual Perception*. Houghton Mifflin.
Gopnik, A., & Schwitzgebel, E. (1998). Whose Concepts Are They, Anyway? The Role of Philosophical Intuition in Empirical Psychology. In M. R. DePaul & W. Ramsey (Eds.), *Rethinking Intuition* (pp. 75–91). Rowman & Littlefield.
Hamblin, C. (1958). Questions. *The Australasian Journal of Philosophy, 36*, 159–168.
Harrah, D. (1985). A logic of message and reply. *Synthese, 63*, 275–294.
Harrah, D. (1987). Hintikka's theory of questions. In R. Bogdan (Ed.), *Jaakko Hintikka* (pp. 199–213). D. Reidel.
Harrah, D. (2002). The Logic of Questions. In D. M. Gabbay & F. Guenthner (Eds.), *Handbook of Philosophical Logic Vol. 8* (pp. 1–60). Kluwer.
Hintikka, J. (1975). Answers to Questions. In H. Hiż (Ed.), *Questions* (pp. 279–300). D. Reidel.

Hintikka, J. (1992). The Interrogative Model of Inquiry as a General Theory of Argumentation. *Communication and Cognition, 25*, 221–242.
Hintikka, J., & Hintikka, M. B. (1989). Questioning as a Philosophical Method. In J. Hintikka & M. B. Hintikka (Eds.), *The Logic of Epistemology and the Epistemology of Logic. Selected Essays* (*Synthese Library* vol. 200) (pp. 215–233). Kluwer.
Hiż, H. (1962). Questions and Answers. *Journal of Philosophy, 59*(10), 253–265.
Hiż, H. (1978). Difficult Questions. In H. Hiż (Ed.), *Questions* (pp. 211–226). D. Reidel.
Ingarden, R. (1925). *Essentiale Fragen. Ein Beitrag zum Problem des Wesens*. Max Niemeyer.
Ingarden, R. (1964). *Der Streit um die Existenz der Welt* (*Vol. 1: Existenzialontologie*). Tübingen: Max Niemeyer. English translation: *Controversy Over the Existence of the World* (Vol. I). Peter Lang, 2013.
Kubiński, T. (1967). Some Observations about a Notion of Incomplete Answer. *Studia Logica, 21*, 39–42.
Kubiński, T. (1980). *An Outline of the Logical Theory of Questions*. Akademie Verlag.
Lewis, D., & Lewis, S. (1975). Review of R. Olson and A. Paul, 'Contemporary Philosophy in Scandinavia'. *Theoria, 41*, 39–60.
Nado, J. (2014). Philosophical Expertise. *Philosophy Compass, 9*, 631–641.
Noë, A. (2012). *Varieties of Presence*. Harvard University Press.
Prior, M., & Prior, A. (1955). Erotetic Logic. *Philosophical Review, 64*, 43–59.
Pust, J. (2019). Intuition. In E. Zalta (Ed.), *The Stanford Encyclopedia of Philosophy* (Summer 2019 Edition). https://plato.stanford.edu/archives/sum2019/entries/intuition/.
Rescher, N. (2000). *Nature and Understanding*. Oxford University Press.
Schaffer, J. (2007). Knowing the Answer. *Philosophy and Phenomenological Research, LXXV*(2), 383–403.
Schellenberg, S. (2011). Perceptual Content Defended. *Noûs, 45*(4), 714–750.
Siegel, S. (2006). Which Properties Are Represented in Perception. In T. S. Gendler & J. Hawthorne (Eds.), *Perceptual Experience* (pp. 481–503). Oxford University Press.
Siegel, S. (2010). Do Experiences Have Contents? In B. Nanay (Ed.), *Perceiving the World* (pp. 333–368). Oxford University Press.
Siegel, S. (2016). The Contents of Perception, In E. Zalta (ed.) *The Stanford Encyclopedia of Philosophy* (Winter 2016 Edition). https://plato.stanford.edu/archives/win2016/entries/perception-contents/

Sosa, E. (2007). Experimental Philosophy and Philosophical Intuition. *Philosophical Studies, 132*(1), 99–107.

Szaniawski, K. (1973). Questions and Their Pragmatic Value. In R. Bogdan & I. Niiniluoto (Eds.), *Logic, Language, and Probability* (pp. 121–123). D. Reidel.

Wiśniewski, A. (1994). Erotetic Implications. *Journal of Philosophical Logic, 23*, 173–195.

Wiśniewski, A. (1995). *The Posing of Questions: Logical Foundations of Erotetic Inferences*. Kluwer.

Wittgenstein, L. (1922). *Tractatus logico-philosophicus*. Kegan Paul. Quotations from: L. Wittgenstein, *Tractatus Logico-Philosophicus*. Mineola, NY: Dover Publications 1999.

4

Philosophy as Recognition: Thinking from Inside the Cave

4.1 Assessing the Current State of the Investigation

I began the book by elaborating on the issue of comparability between what can provisionally be bundled up into the umbrella concept of *philosophical beliefs*, that is, views articulated by professional philosophers, and beliefs that seem, at least on the surface, related to philosophical issues, yet such that a wider group of people dubbed "non-philosophers" can hold or be disposed to endorse them (if asked).

One of the rationales for the project pursued here is the challenge that the emergence of Xphi poses to more traditional ways of philosophizing. For example, as a result of the famous studies of Nichols et al. (2003), it turns out, first, that traditional epistemologists' tacit assumptions as to what is intuitively regarded as knowledge are not as pervasive in the wider population as had been assumed; and second, that one's allegedly intuitive understanding of knowledge hinges on such variables as one's cultural background, gender, and even socioeconomic status. This naturally gives rise to doubts as to what professional philosophers are in fact investigating; what is it that they specialize in, as distinguished from, for

K. Werner, *The Embodied Philosopher*, New Directions in Philosophy and Cognitive Science, https://doi.org/10.1007/978-3-030-79964-9_4

example, historians specializing in, say, the rise and fall of the Polish–Lithuanian Commonwealth or Victorian England. Arguably, it makes no sense whatsoever to examine the historians' competence vis-à-vis what the general public knows. Analogously, if there is no "thing" that philosophers know better than others, what is the rationale for their work?

I argue that the key to understanding philosophical competence is to have a proper grasp of how selected parts or aspects of the world are targeted by acts of philosophical problematization (questioning). But how does it come about that we are able to problematize anything? Chapter 3 pointed to a specific cognitive capacity called erotetic intuition. It was first figuratively characterized as the capacity to directly "see" problems in a given domain even when there is no clear-cut logical sequence of steps—instructions or other operations—that produce them. Yet this *presence of problems* does not come cheaply; we must earn it, so to speak, and Chap. 5 shall provide a more detailed, albeit grossly simplified, account of how questioners bring this about.

At this point the question is, what specific acts of problematization set the stage for the *pursuit of philosophizing*? What is the specifically philosophical attitude that a subject—a living creature capable of intellectual endeavor—can adopt with regard to her environment? This chapter seeks to pinpoint what it is that makes certain problems or questions *philosophical*.

Yet instead of merely listing questions that would be regarded as philosophical, for example, questions about truth or knowledge, I propose to focus on a specific way of exercising erotetic intuition. The puzzle is therefore what makes certain ways of exercising of one's erotetic intuition philosophical *for* the subject in question. Crucially, it is irrelevant at this point whether the subject is a "professional" philosopher.

4.2 Russell and Carnap on the Troubling Nature of Philosophizing

Arguably there is a common wisdom according to which philosophers are skilled in taking the world we experience as a whole and in problematizing the most general issues which may be transparent, unnoticed, from the everyday perspective as well as the perspective of the sciences, focused as they are on empirically delineated segments or aspects of reality. For example, philosophers do not ask whether this or that empirical investigation can result in a piece of new knowledge, but under what conditions *empirical knowledge as such* is gleaned.

Now, at some point there was consensus among at least some philosophers (those very critical about their own discipline) that this level of generality and abstractness is also dangerous. Bertrand Russell expressed his concerns to this effect in "The Problems of Philosophy," among other works. Let me make use of several extensive quotes from this brilliant work, as they clearly illustrate the line of thinking about philosophy that I wish to follow. At the same time, Russell's book also highlights a certain subtle tension which is also at the core of the conception I wish to roll out here.

Russell expresses his reservation toward philosophers, particularly "metaphysicians," who would approach the world *en bloc* and strive to determine its ultimate, true nature:

> Most philosophers—or, at any rate, very many—profess to be able to prove, by a priori metaphysical reasoning, such things as the fundamental dogmas of religion, the essential rationality of the universe, the illusoriness of matter (…). This hope, I believe, is vain. It would seem that knowledge concerning the universe as a whole is not to be obtained by metaphysicians, and that the proposed proofs that, in virtue of the laws of logic, such and such things must exist and such and such others cannot, are not capable of surviving a critical scrutiny. (Russell 1912/2004, p. 99)

Aiming at the world *as a whole* is a non-starter, according to Russell. On the other hand, however, he himself points out that philosophy is not concerned with well-delineated fragments of reality, but with certain "ultimate questions." He starts his book with the following words:

> Is there any knowledge in the world which is so certain that no reasonable man could doubt it? This question, which at first sight might not seem difficult, is really one of the most difficult that can be asked. When we have realised the obstacles in the way of a straightforward and confident answer,

> we shall be well launched on the study of philosophy—for philosophy is merely the attempt to answer such ultimate questions, not carelessly and dogmatically, as we do in ordinary life and even in the sciences, but critically, after exploring all that makes such questions puzzling, and after realizing all the vagueness and confusion that underlie our ordinary ideas. (ibid., p. 1)

He then adds elsewhere that for a "man who has no tincture of philosophy" "the world tends to become definite, finite, obvious; common objects rouse no questions, and unfamiliar possibilities are contemptuously rejected" (ibid., pp. 109–110).

Russell is therefore of the opinion that philosophical reflection differs from everyday thinking in crucial ways. Namely, philosophy is due to a change of one's habitual, ordinary attitude toward the world which is, supposedly, marked by presuppositions, biases, customs, ideologies, and so on. Whether or not this claim is empirically confirmed (when articulated as a testable hypothesis, it can be taken on by members of the Xphi camp), the crucial conceptual point is this: There is *a distinction*, also drawn by Husserl and Carnap, as we shall see later on, between a natural attitude toward the world and toward thinking and a specifically *philosophical* attitude. I keep it open at this point what these two conceptually distinguished attitudes actually stand for. Nonetheless, let us provisionally assume that some distinction of this kind is the case. As for attitudes themselves, I assume here that the term is understandable, that is, that we know what having an attitude of any sort means: it refers either to a mental state directed toward something or to a disposition to cognize something thus and so (more on the two views of attitudes, see Machery 2016). The next question is this: What kind of "machinery"—conceptual, logical, or intellectual in the broadest sense—is responsible for one's *change* of attitude.

Russell is explicit about one thing. He says that there is no special intellectual faculty that is specific to philosophy; hence no philosophical intuition proper, either. Noteworthily, when it comes to more recent meta-philosophical investigations, Machery (2017) is of a very similar opinion. Yet, the conclusion drawn by Russell is quite far-reaching, namely that "the results obtained by philosophy are not radically different from those obtained from science" (ibid., p. 105).

If this is the case, however, how can we account for the link, explicitly mentioned by Russell, between "the most everyday things" on the one side, and the "possibilities which enlarge our thought" on the other? Is the step from the former to the latter just another exercise of our usual intellectual faculties? Here is the fragment where the two mentioned notions are introduced:

> As soon as we begin to philosophise, (…) we find (…) that even the most everyday things lead to problems to which only very incomplete answers can be given. Philosophy, though unable to tell us with certainty what is the true answer to the doubts which it raises, is able to suggest many possibilities which enlarge our thoughts and free them from the tyranny of custom. Thus, while diminishing our feeling of certainty as to what things are, it greatly increases our knowledge as to what they may be; it removes the somewhat arrogant dogmatism of those who have never travelled into the region of liberating doubt, and it keeps alive our sense of wonder by showing familiar things in an unfamiliar aspect. (ibid., p. 110)

How is it possible that "the most everyday things lead to problems"? What is the *mechanism* responsible for "diminishing our feeling of certainty as to what things are," for the "knowledge as to what they may be," thus for the move from "the most everyday things" to the realm of possible worlds? How does it happen that "familiar things" uncover their "unfamiliar aspects"? *How* does philosophy manage to do all these things? Russell answers these concerns by saying that philosophy "examines critically the principles employed in science and in daily life; it searches out any inconsistencies there may be in these principles" (ibid., p. 105).

I argue that this is not a satisfactory answer. Criticism is also part of everyday thinking; there is no a priori reason to assume that philosophers are more critical than others. On the contrary, empirical findings show them to be as biased as anybody else (see Chap. 2). Perhaps the point is not in the scope or intensity of their individual (personal) criticism, but in philosophical criticism's *unique approach* toward various everyday things: one that is attributed not to persons but rather to roles, thus, to use Fine's (1982) terms, to persons *qua philosophers*. A similar remark can be made with regard to Russell's idea of tracing and eliminating inconsistencies. Which inconsistencies, exactly? Our everyday thinking is likely full of inconsistent views, but is philosophy supposed to elicit all of them, or just those that are philosophically significant? Yet how are those philosophically significant views differentiated from those significant in other fields of inquiry or in everyday life? The answer I wish to flag in this chapter can actually be read from Russell's deliberations and his overall standpoint (as we shall see later on), yet it is not articulated explicitly there.

I move on now to provide a working characterization of the contents attributable to philosophy. Interestingly enough, their specificity has been acknowledged most by those who continued Russell's criticism of the "metaphysicians" and came up with even more far-reaching claims with respect to the very stance of philosophy. In this group Rudolf Carnap and Ludwig Wittgenstein likely made the most explicit suggestions that philosophical contents could *not* be products of ordinary processes of perception, categorization, or reasoning. This is because, in their view, apprehending the world the way philosophy does is *not just another step* within the hierarchy of generality, but a *major transgression* of conceptual schemata based on ordinary natural-generic concepts, and a movement either toward nonsensical pseudo-problems (as Carnap, and in some other way Wittgenstein, too, claimed) or toward something more fundamental which underlines the empirical domain and the pursuit of science (for lack of a better word, we can call this latter group *defenders*, e.g., Husserl; surprisingly Carnap belongs here, too, yet interpreted from a different angle—see below).

Carnap's (1950) distinction between internal questions and external questions addresses the specificity of philosophizing directly. I shall discuss this distinction in greater detail further on, for now let me provide a short introductory rendering. First, Carnap introduced the notion of linguistic framework which is more or less equivalent to the earlier idea of conceptual apparatus set forth by Kazimierz Ajdukiewicz (1934). This is, roughly, a set of "ways of speaking," thus expressions equipped with references and strict rules which determine which expressions are used properly in specific contexts. According to Ajdukiewicz, each conceptual apparatus can be represented by (but not reduced to) a setup of "meaning rules," which at the end of the day are rules of behavior. If, for example, one refuses to accept the sentence *śnieg pada* when faced with white stuff everyone recognizes as snow falling from the sky, one is not making a false statement in Polish, but, at a more primitive level, one is not obeying an empirical meaning rule establishing, together with a large number of similar rules, the apparatus of the Polish language. Truth and falsity emerge as an issue only when we already have a particular language set up (a conviction that was quite common in the Lvov-Warsaw school and made famous, of course, by Alfred Tarski; see his collected papers from 1923 to 1938, Tarski 1983).

Now, according to Carnap, in addition to propositions, within every linguistic framework (conceptual apparatus) there are also questions. Carnap especially targets "questions of existence" in his analysis, but the idea at stake could be generalized. He starts by considering the usual "thing language" in which we can sensibly form questions such as "is there a book on my desk?" or "are unicorns real or imaginary?" which "are to be answered by empirical investigations. Results of observations are evaluated according to certain rules as confirming or disconfirming evidence for possible answers" (Carnap 1950, p. 21). He then adds:

> The concept of reality occurring in these internal questions is an empirical scientific non-metaphysical concept. To recognize something as a real thing or event means to succeed in incorporating it into the system of things at a particular space-time position so that it fits together with the other things as real, according to the rules of the framework. (ibid., p. 22)

Being an internal question means being *internal to* the fixed linguistic framework. In this group we find questions formulated by ordinary people as well as scientists, aside from the technicalities introduced by the latter. Now, Carnap says:

> From these questions we must distinguish the external question of the reality of the thing world itself. In contrast to the former questions, this question is raised neither by the man in the street nor by scientists, but only by philosophers. Realists give an affirmative answer, subjective idealists a nega-

> tive one, and the controversy goes on for centuries without ever being solved. And it cannot be solved because it is framed in a wrong way. To be real in the scientific sense means to be an element of the system; hence this concept cannot be meaningfully applied to the system itself. (ibid.)

Carnap's distinction is not unproblematic for various reasons to be unpacked later in greater detail. What is critical at this point is that unlike Russell, Carnap seems to suggest that there *is* a philosophy-specific intellectual faculty of questioning, whose allegedly failed character is one thing, but whose *distinctness* is another. Let me put it this way: independently of whether or not we agree with Carnap, there must be, on the basis of his view, a fundamental difference between the act of questioning, thus the act of exercising one's erotetic intuition that generates internal questions and the act of questioning generating external questions. Whether the philosophical questions are a serious misstep (as he argued) or a virtue (as a defender would argue), one thing is clear—they are *not* ingredients of our ordinary intellectual toolbox. So we have two kinds of question-, and more generally, content-generators. I shall follow the latter idea.

Against Carnap (and Wittgenstein 1922, too; see 6.5 and 6.51) I provisionally assume, for the sake of this investigation, that there are at least some plausible philosophical (external) questions. Otherwise this investigation into the nature of philosophical intuition would be in principle in vain. What this plausibility amounts to is going to be unpacked in due course.

4.3 The Apparent Banality of Philosophical Questions (and Beliefs)

It is one thing to ask whether philosophizing produces nonsensical utterances, and another to ask whether it can produce valuable or insightful utterances, that is to say questions and answers that draw our attention to something new, interesting, challenging, useful, and so on. As I already mentioned, I assume that external questions can make sense (to be unpacked later on). At the same time, I wish to raise the issue of their apparent *banality*.

This is an extremely subtle issue. Philosophical content is not generated by using some peculiar technical language, but—as I will argue—through a specific attitude toward the world, which nevertheless generally uses the expressive devices of natural language. Of course, philosophical language has established its own technicalities, but, as a rule, philosophical content can be successfully articulated using common expressions. The price to be paid is the apparent banality of virtually all attempts to articulate philosophical concerns.

I am talking about the banality manifesting itself when one looks at philosophical problems and claims simply as ordinary expressions in natural language. These claims appear banal or trivial when one looks at them *from within* everyday experience and from within the conceptual frameworks designed to capture that experience. To put it in yet a different manner: philosophical questions may appear trivial if one looks at them *as if they were* internal questions. Thus, if we adopt the evaluation criteria applied to non-philosophical intellectual efforts directed at a given fragment or aspect of the world, then the outcomes of philosophical efforts seem not to alter anything, that is, not to deliver any new knowledge.

Let us consider an example. Maralee Harrell begins her (2016) introduction to philosophy in the following way:

> Many Introduction to Philosophy textbooks begin with an admonition that people generally don't understand what philosophy is. I don't find this to be the case, though. On the first day of class, I always ask my students to think of questions that they think are particularly philosophical; and they always do a great job (…). Usually the list goes something like this:
>
> Does God exist?
> What is human nature?
> What can we know? (…)
> What is real?
> What is consciousness?
> Is truth relative? (Harrell 2016, pp. 3–4; see e.g., Horner & Westacott 2000 in the same spirit)

As already hinted at in Chap. 2, nothing a priori makes the question "What is real?" a philosophical one, while on the level of common interpretation the answer to it may be trivial: the table at which I am sitting is real, while the dinosaur I see when wearing a virtual reality headset is not. If the philosopher reads the same question differently, for example in the context of the *realism vs. idealism* dispute, then there must be an additional factor which, so to speak, *de-banalizes* the question. The same goes for the question regarding the relativity of truth, or, more generally, the nature of true propositions (judgments). The answer to the question of what it means for the proposition "Snow is white" to be true is seemingly trivial for anyone who has ever seen snow, while the issue of the relativity of truth is immediately resolved by anyone who can see the difference between the propositions "2 + 3 = 5" and "Zimerman is a better Chopin player than Pollini." Indeed, there are relative truths, although not all truths are relative (*ergo*: what is there to "philosophize" about?). Moreover, even if the question is not apparently trivial, it is by no means obvious that it is philosophical. The question of God, cited by Harrell, is philosophical if "God" refers, for example, to the specific being that Anselm of Canterbury had in mind in his ontological proof, but a Christian or Jewish theologian likely understands this utterance differently. The same applies to the question of consciousness and how the latter is operationalized in cognitive science.

To take a different example, note that some of the most puzzling points made by Wittgenstein in his *Tractatus* are apparent trivialities:

> 5.621 The world and life are one.
>
> 5.63 I am my world. (The microcosm.) (Wittgenstein 1922/1999, p. 89)

Yet, we suspect that there is something laid down in these claims that renders them non-trivial. So putting together all the things discussed so far we must seek to understand the nature of this "de-banalizing" factor (if any), which must partake in the machinery of erotetic intuition used by any subject engaged in the pursuit of philosophizing, so that it can legitimize philosophy as an intellectual effort.

Of course, we must also take into account and closely consider the option that no such de-banalizing factor exists, thus that some traditional philosophical (particularly metaphysical) claims are trivial at face value and turn out to be either unanswerable or even nonsensical under logical scrutiny. We shall discuss the meta-philosophical critique advanced by Carnap (1950) and more recently (in a different framing) by Machery (2017) in due time.

4.4 The Idea of a Radical Beginning

Seeking a de-banalizing factor, I believe it is worthwhile to look at philosophical schools which, contrary to what Harrell (and Searle 1998, too) assumed, *did not* consider the question of reality, truth, or similar concerns as the starting point, but demanded going even further, digging down to a level of reflection where thought could gain some neutral footing with respect to these issues. For only then, according to this approach, all those further puzzles could be taken on, including the question of reality, mind-dependence, truth, and knowledge. So, I will be offering a working reflection on the philosophies of *the radical beginning*.

The idea of a radical beginning comes from Descartes, and in the twentieth century it was explicitly formulated by Husserl, who consciously referred to the author of the "Meditations on First Philosophy" and harked back to his first-person, meditative method of philosophizing. Husserl emphasizes that the quest for a radical starting point is individual; it begins with a specific kind of self-knowledge, when I myself become the center or point of reference in relation to which all the manifest and scientific images of the world, as Sellars (1956) put it, are to be situated. In "Cartesian Meditations" we read:

> First, anyone who seriously intends to become a philosopher must "once in his life" withdraw into himself and attempt, within himself, to overthrow and build anew all the sciences that, up to then, he has been accepting. Philosophy (…) is the philosophizer's quite personal affair. It must arise as his wisdom, as his self-acquired knowledge tending toward universality, a knowledge for which he can answer from the beginning, and at each step,

> by virtue of his own absolute insights. If I have decided to live with this as my aim (...) I have thereby chosen to begin in absolute poverty, with an absolute lack of knowledge. (Husserl 1950/1982, p. 2)

In Husserl's opinion, this step toward radical self-knowledge is linked to a certain attitude toward the world which is not possible in everyday, ordinary approaches, characterized as the natural attitude. In *Ideas I* we read:

> By my seeing, touching, hearing, and so forth, and in the different modes of sensuous perception, corporeal physical things with some spatial distribution or other are *simply there for me, "on hand"* in the literal or the figurative sense, whether or not I am particularly heedful of them and busied with them in my considering, thinking, feeling, or willing. (Husserl 1913/1983, p. 51)

Earlier in the same book he writes:

> Natural cognition begins with experiences and remains *within* experience. In the theoretical attitude which we call the *"natural"* <*theoretical attitude*> the collective horizon of possible investigations is therefore designated with *one* word: It is the *world.* Accordingly, the sciences of this original attitude are, in their entirety, sciences of the world. (ibid., p. 5)

The key, therefore, is not so much that the existence of the world is naturally assumed, not to mention being deliberately endorsed as part of the natural attitude. Rather, in the natural attitude the very presence of the world, as it is experienced, appears as something so obvious ("at hand") as not to inspire reflection at all—it is *not* problematized in thinking. The world is just *there*, present to me. Bracketing the natural attitude, postulated by Husserl—the epoche—is therefore a condition which makes the said problematization (not denial) possible, even before any further reflective steps have been taken.

> Instead of remaining in this [natural] attitude, we propose to alter it radically. (ibid., p. 57)

> *We put out of* action *the general positing which belongs to the essence of the natural* attitude; we parenthesize everything which that positing encompasses with respect to being: *thus the whole natural world* which is continually "there for us", "on hand," and which will always remain there according to consciousness as an "actuality" even if we choose to parenthesize it.
>
> If I do that, as I can with complete freedom, then I am *not negating* this "world" as though I were a sophist; I am *not doubting its factual being* as though I were a skeptic; rather I am experiencing the "phenomenological" ἐποχή which also *completely shuts me off from any judgment about spatiotemporal factual being.* (ibid., p. 61)

The crucial thing is this: if the epoche is neither negating nor doubting the experienced world, it is all about *taking the world solely as it manifests itself*—taking it in its relation to the subject, thus *qua experienced*, yet without making it subordinate to the subject (which is a metaphysical claim, bracketed alongside all other metaphysical claims).

Interestingly, the notion of anchoring science in what is given *as given* was also the premise of Carnap's (1928/1967) Aufbau project, that is, the idea of creating a coherent system of concepts (a constitutive system, as Carnap called it) founded on basic notions, which would then be harnessed within a suitable logical machinery to enable the unity of the sciences, that is, produce a single coherent scientific image of the world. This can be seen most clearly when Carnap selects the most appropriate *base* for his system. He finds it in subjective experiences taken not as the creations of an actual psychophysical subject, but as neutral atoms of cognition: they are to serve *as* given, and not a result of some kind of relationship to a supposed reality. In fact, Carnap directly references an analogous effort by Husserl. He writes:

> The autopsychological basis is also called solipsistic. We do not thereby subscribe to the solipsistic view that only one subject and its experiences are real, while the other subjects are nonreal. The differentiation between real and nonreal objects does not stand at the beginning of the constructional system. As far as the basis is concerned, we do not make a distinction between experiences which subsequent constructions allow us to differentiate into perceptions, hallucinations, dreams, etc. This differentiation and thus the distinction between real and nonreal objects occurs only at a relatively advanced constructional level (…). At the beginning of the system,

> the experiences must simply be taken as they occur. We shall not claim reality or nonreality in connection with these experiences; rather, these claims will be "bracketed" (i.e., we will exercise the phenomenological "withholding of judgment", ἐποχή, in Husserl's sense). (Carnap 1928/1967, p. 101)

Yet, it should be clear that this specific neutrality obtained by the exercise of "bracketing" *cannot be part of the natural* attitude. A major reason for that is the fact that in Carnap's approach the givenness "does not presuppose somebody or something to whom the given is given" (ibid., p. 102). Further on we read:

> At the outset [i.e., at the base level of the reduction], we can speak neither of other subjects nor of the self. Both of them are constructed simultaneously at a higher level. (ibid., pp. 103–104)

When it comes to the everyday view of the world, however, it is difficult to imagine grasping what is given only within the bounds of this "givenness," so to speak. The natural attitude is determined, *inter alia*, by the distinction between subject and object, while the radical starting point, as Carnap aptly put it, should lie one step ahead of this distinction. *Egocentricity*, as the author of "The Logical Structure of the World" calls it, is a type of construct, not something given.

Now, I am not ruling whether the search for a radical starting point in philosophy can be successful (whatever being successful means in this context). I only claim that there is a take from the fact that such an attempt is at the heart of otherwise very different systems such as those developed by Husserl and Carnap. First of all, in addition to the fact that philosophical claims are closely correlated with appropriate questions, which was elicited in Chap. 2, it is now also apparent that these questions have to result from a very peculiar change of attitude, thus from a very peculiar exercise of one's erotetic intuition. In short: it is virtually impossible to produce a philosophical problem, at least from the perspective shared by Husserl and Carnap, without bracketing the natural attitude. This radical meta-philosophical position—somewhat abandoned these days and openly re-affirmed in this book—casts doubt on the procedure of questioning people about philosophical matters without teaching them how to exercise bracketing or something similar (which is, in principle, a thinkable scenario provided that Xphi enriches its methods; see Andow 2016). As announced, the problem with this procedure is *not* that people give amateur answers; the real issue is that they may be giving answers to different questions.

Secondly, if the idea of a radical beginning is at the heart of otherwise very different systems such as those developed by Husserl and Carnap, it is at least a considerable option that the bracketing effected by both thinkers is an instantiation of a more general intellectual pursuit—or a pattern thereof—standing at the heart of philosophizing as such (by the way, we should not overemphasize the differences between Husserl and Carnap, for although they would later become—which is not necessarily a good thing—sharply distinguished as founding fathers of two different currents in contemporary philosophy, their points of departure had a lot in common). I shall risk the hypothesis that the latter is true. The common intellectual source of philosophizing, I argue, is what I dub the *recognition of one's epistemic position*. And this specific intellectual act is the sought de-banalizing factor which renders the mentioned seemingly trivial questions non-trivial. This act, not an attitude itself, serves as the driving force behind the change of attitudes—from the natural to the philosophical.

Crucially, the recognition in question is first and foremost a certain experience. This aspect was diligently developed by Husserl and somewhat ignored by Carnap. The next section provides a brief characterization of its content.

4.5 "The World Is My World" Beyond Wittgenstein: The Recognition of One's Epistemic Position

To begin this part, let us take the bull by the horns and flag what is likely the biggest triviality a philosopher can turn into a recognition. Here it is: minded creatures face a certain layout, presentation, or image of the world (a "manifest image" as Sellars 1956 called it). Being an *image*, it first of all depends on the creature's cognitive apparatus, and could therefore be different. Secondly, it should be distinguished from what it is an image *of*, that is, the supposed reality in itself.

Philosophers have been aware of this seemingly trivial "fact" for centuries, including especially Kant and phenomenologists, yet in my opinion Tim Crane provides an exceptionally appealing description in simple words:

> The minded creature is one *for which* things are a certain way: the way they are from that creature's perspective. (…) We might express this by saying that a minded creature is *one which has a world*: its world. Its having a perspective consists in its having a world. Having a world is something different from there simply being a world. It is true of the rock or the daffodil that it is part of the world; but it is not true that they have a world. (Crane 2001, p. 4)

Here is Maurice Merleau-Ponty:

> The world is inseparable from the subject, but from a subject which is nothing but a project of the world, and the subject is inseparable from the world, but from a world which the subject itself projects. The subject is a being-in-the-world. (Merleau-Ponty 1962, p. 430)

One of the founding fathers of theoretical biology, Jakob von Uexküll, put it in more technical terms:

> All that a subject perceives becomes his perceptual world and all that he does, his effector world. Perceptual and effector worlds together form a closed unit, the Umwelt. (von Uexküll 1934/2010, p. 6)

Francisco Varela and his collaborators, to whom I shall especially refer in Chap. 5, draw on Merleau-Ponty as well as von Uexküll when they state:

> A phenomenologically inclined cognitive scientist reflecting on the origins of cognition might reason thus: Minds awaken in a world. (…) We reflect on a world that is not made, but found, and yet it is also our structure that

> enables us to reflect upon this world. Thus in reflection we find ourselves in a circle: we are in a world that seems to be there before reflection begins, but that world is not separate from us. (Varela et al. 1991, p. 3)

I will argue here that being aware of the "trivial fact" put into words in the above quotes is the very essence of philosophical competence. *Becoming* aware of it is exactly what shall be referred to as the recognition of one's epistemic position.

According my overview, a subject's epistemic position involves at least the following four factors (this list is by no means exhaustive):

Firstly, the subject's location. Hilary Putnam in known, among many other things, for his rejection of the "God's-eye view," which prevails, as he holds, in modern epistemology. This refers to the epistemic ideal of achieving a view from an " 'Archimedean point'—a point from which we can survey observers as if they were not ourselves, survey them as if we were, so to speak, outside our own skins" (Putnam 1990, p. 17). In an illuminating reading of Kant, Henry Allison (2006) calls this prevailing epistemological view a *theocentric epistemology*, since it seems appropriate for God, but surely not for finite minds. According to this view human knowledge "is to be measured and evaluated in terms of its conformity (or lack thereof) to the norm of a putatively perfect divine knowledge" (Allison 2006, p. 114). In other words, we are supposed to be capable of gaining a view from nowhere, as Nagel (1986) would say.

In opposition to the theocentric ideal of a view from nowhere, being in an epistemic position means first of all being *somewhere*, being in a particular *position* with respect to all other beings and thus having *local* perspective(s). It means living inside the world and dealing with it *from the inside*; it means being saddled with all the limitations imposed by the very fact that we are, as subjects, positioned.

Secondly, the epistemic position of a subject is constituted by the subject's perceptual capacities. Philosophers since the time of Heraclitus have been aware of the fact that our senses provide us only with fractional data; therefore, it seems reasonable to distinguish between the world as it allegedly is in itself and the ways in which it appears to us via perception. Early modern theories of perception, based on Galileo's mathematical physics and Kepler's optics, created a gulf between a mechanistically conceived external reality on the one hand, and the sensible qualities of the world as it is known by perceivers on the other (see Ben-Zeev 1984; Brown 2008; Osler 2008; and Chap. 5). On this basis Descartes and Locke laid down the categorical scaffolding of primary and secondary qualities which—under different descriptions—would later also become a standard for empirical studies in cognitive science (for a critique of the primary-secondary distinction see Putnam 1987).

Being in an epistemic position thus means that local subjects face an appearance of the world that fits their perceptual capacities. The appearance of the world that I know is determined by the capacities of my receptors, by the whole machinery of my neural network, by the bottom-up mechanisms of selection of stimuli, and so on.

Thirdly, besides perceptual constraints that we share with the rest of the animal kingdom, the human epistemic position is also constituted by the intellectual machinery of language and rational thinking. One might argue that the human mind transcends the limits imposed by biological factors, and admittedly that might be the case, yet it does not change the seemingly obvious fact I am highlighting in this section, namely that our worldview is still determined by these intellectual conditions. Searle says in this context that "all intentionality is aspectual. Seeing an object from a point of view, for example, is seeing it under certain aspects and not others. In this sense, all seeing is 'seeing as'. And what goes for seeing goes for all forms of intentionality, conscious and unconscious" (Searle 1992, p. 131). We are thus accustomed to imposing, as Kant already taught us, a categorical "architecture" on the world so that we see objects and properties, processes and events on a timescale; we see relations, causes and effects, and so on.

Hence, being in an epistemic position means for humans (likely some other animals, too, in a more "primitive" manner) that they organize their worlds by imposing categorical and conceptual frameworks on them.

Finally, the subject who recognizes her local position along with all its constraints is the one who says "I," and this very expression opens up a new dimension of epistemic position: the "fact" that "the world is my world" (Wittgenstein's "Tractatus" 5.641), meaning that all facts are, in a sense, *for me*. Recently Dan Zahavi and Uriah Kriegel have aptly called this feature "for-me-ness." They write:

> Arguably, for every possible experience we have, each of us can say: whatever it is like for me to have the experience, it is for me that it is like that to have it. What-it-is-like-ness is properly speaking what-it-is-like-for-me-ness. (Zahavi and Kriegel 2016, p. 36)

Thus, being in an epistemic position means that the world I am *in* is always the world *for me*. I am a constant and necessary point of reference, as it were, for whatever I could call a world.

Now, recognizing my epistemic position is my becoming aware of the "facts" described above. It is therefore about me recognizing everything that happens ("is the case," as Wittgenstein puts it) precisely as that which happens around *me* and is somehow tailored to *my* cognitive measure. It is about pointing out that the whole world *as it appears* to *me*, the world qua experienced, is a phenomenon, and thus a certain presentation. Finally the "me" part is also a certain (self)presentation. Recall Carnap—egocentricity is a construct (here—a presentation), too.

Partial recognitions of this type sometimes occur in everyday life, when, for example, we discover that our senses sometimes deceive us. Cognitive psychology, which examines the contextual dependence of perception, conceptualization, memory, and so on, also provides relevant examples. However, one of the special features of the recognition of one's epistemic position, as understood here, is its holistic character; that is, the *whole* empirical domain is apprehended as that which appears—as given, in Carnap's (1928/1967) sense. It is therefore easy to counter faulty perceptions of shapes under certain circumstances or erroneous categorization within a certain context with adequate perception and adequate categorization in standard conditions, yet we don't have access to anything with which the empirical domain *as a whole* could be compared. This wholeness, then, is something specific which seems to require *more* than just the summation of specific cases; namely, a qualitatively new step, resulting from a kind of special reflection or meditation of the kind undertaken by Descartes, Husserl, or Wittgenstein.

Having articulated all these trivialities, recognizing one's epistemic position opens up the prospect of *a new kind of generality*—a generality which isn't only the result of adding up local similarities and differences between perceptions, features, or phenomena, but one that allows the whole world qua experienced to be grasped in a single act of the intellect. If we approach this act from the perspective of Chudnoff's (2011) perceptualist account of intuition, it is clear that we aren't dealing with a belief of any kind, but certainly some content is just presented to the subject so that the subject can actually "see" or experience the world and herself in the world *anew*, even though nothing has changed "objectively."

Therefore, the recognition of one's epistemic position could be considered as an experience of some specific sort, much like aspectual seeing in Wittgenstein's later writings:

> The expression of a change of aspect is the expression of a *new* perception and at the same time of the perception's being unchanged. (Wittgenstein 1953, p. 196)

When I recognize a rabbit in the famous duck-rabbit illustration invoked in "Philosophical Investigation," "objectively" nothing has changed in the world, yet the whole world is different *to me*; from now on it is *a world with a rabbit.*

The recognition of one's epistemic position is thus akin to aspectual seeing: it enables one to see the *whole* world under a new aspect. As such, it puts the whole world in a new frame, so to speak, while at the same time it may seem that nothing changes. It's just like changing "the limits" of the world, not the world itself, as articulated in the following fragment from Wittgenstein's *Tractatus*:

> 6.43 If good or bad willing changes the world, it can only change the limits of the world, not the facts; not the things that can be expressed in language.
>
> In brief, the world must thereby become quite another. It must so to speak wax or wane as a whole. (Wittgenstein 1922/1999, p. 106)

Arguably, this is the reason behind the apparent triviality or banality of philosophical questions and answers. Namely, they seem such when they are supposed to refer to some facts; meanwhile they refer to what Wittgenstein calls "the limit" of the domain of facts. As in aspectual seeing, "philosophical seeing," if I can put it this way, changes nothing in particular, yet it renders the whole domain of experience and thinking different. So, in a way, while changing nothing in particular, it changes everything.

The key thing, already flagged above, is the emergence of a new kind of generality—one that radically transcends the ascending hierarchy of natural kinds and demands a different sort of reflection (see Kołodziejczyk 2006 in this context). Arguably the progenitor of this radical move was Plato. First think of his famous metaphor of the cave in light of what has been put forward so far: all the cave-dwellers were likely able to draw conclusions from their perceptions and exercise their erotetic intuition in the way that suited their epistemic position—the position of cave-dwellers—yet the one who decided to leave must have first taken aim not at *this or that* part of the world as presented to him, but at this presentation *as a whole*. Otherwise *leaving the cave* would not have been an option that could ever occur to him. And the very idea of leaving refers not so much to an effective repositioning of oneself, for this would clearly be a non-starter, at least given what we know today about the cognitive constraints determining our epistemic position, but, in my reading, to prospects for a new kind of thinking, questioning, and answering. The latter change nothing particular inside the cave while changing the entirety of our grasp of the world—the cave—in which we are placed.

Plato was quite open about this new discourse, this new kind of generality, especially in *Phaedo* (96a–102a) where he described what is now known in the literature as the Second Voyage (see Reale 1990; and Kraemer 1990 for a different reading)

In my view, this act of *going beyond local generality*, toward what Wittgenstein calls the "limits," along with the "feeling of the world a limited whole," has to involve a specific act of self-knowledge: a presentation of oneself as the one *to whom* the world as a whole appears and, at the same time, a presentation of the world as *that which* appears. Wittgenstein puts it aptly:

> 5.62 (…) In fact what solipsism means, is quite correct, only it cannot be said, but it shows itself.
>
> (…)

> 5.641 There is therefore really a sense in which in philosophy we can talk of a non-psychological *I*. The *I* occurs in philosophy through the fact that the "world is my world".
>
> The philosophical *I* is not the man, not the human body or the human soul of which psychology treats, but the metaphysical subject, the limit—not a part of the world. (ibid., p. 90)

In this light the already cited "trivialities" become expressions of a *philosophical* thought:

> 5.621 The world and life are one.
>
> 5.63 I am my world. (The microcosm.). (ibid., p. 89)

Therefore, the sought de-banalizing factor is the specific act, and at the same time the specific experience of recognizing oneself as "the limit—not a part of the world," and of the world as "my world."

I propose that it is the recognition of one's epistemic position that distinguishes the pursuit of philosophizing from other intellectual endeavors. In other words, the recognition of the philosophizing subject's epistemic position, brought about by the subject's erotetic intuition and then spelled out in the form of specific desiderata, is laid down, whether it is made explicit or not, in the entire philosophical discourse. Philosophical problems and conceptions, as they are worked out in academic philosophy, are relatively stable products of these specific acts of erotetic intuition (in the sense of Twardowski's 1911 *act vs. product* distinction). As products they can be analyzed independently of their origin. This creates an abstract (actually, *abstracted*) realm of philosophical reflection, articulated in a certain technical vocabulary and captured by a particular division of labor, as any other intellectual pursuit these days, yet *not* defined or constituted by these latter factors. Instead, let me reiterate, it is constituted, made what it is, by its being grounded in the philosophizing subject's recognition of her epistemic position, thus a very unique, although also quite troubling, use of her erotetic intuition.

4.6 Philosophical Desiderata: Toward the De-banalization of Philosophical Utterances

Let us now reread some of the questions cited above, beginning with Harrell's (2016) "what can we know?" According Chap. 3, a desideratum should begin with "bring me to know...," but in order to avoid a vicious circle and the need for painstaking distinctions between knowledge, meta-knowledge, and so on, let me put this in the following way: "Bring to my attention/understanding *x* that I can know something about." This, however, is ambiguous, which elicits the ambiguity of the question itself. First, it can be understood in a way that clearly has nothing to do with philosophy, such as "bring to my attention/understanding *x* that I'm allowed to know as opposed to *y* that I am not allowed to know in virtue of state law." The latter formulation applies to cases such as, say, my knowledge of bills signed into law by the Polish president: of course, usually I can know them, meaning I am entitled to read them, as opposed to classified documents which I cannot read.

So the desideratum must make it clear that a different, more fundamental *knowability* is at stake. If we take something like "bring to my attention/understanding *x* that I can know as opposed to *y* such that I cannot know from my current standpoint or perspective," this may have a philosophical meaning, but not necessarily. It can also refer to the fact that from my *current perspective*, for example, I cannot know whether my bank is open or closed; in other words I would have to change my position and drive there or call somebody to find out.

So, how do we manage to arrive at desiderata that seek to abstract from all such *accidental* constraints; thus at something like "bring to my attention/understanding *x* which is a kind of entity I am *in principle* capable of knowing given all my usual perceptual and intellectual tools, as opposed to kinds of entities that I cannot know in the relevant sense." This kind of desideratum, however, can hardly come to an unprepared mind, meaning those who have not *already* ventured in a specific kind of intellectual activity. To be clear, this group does not have to be identical to the institutionally delineated camp of "professional philosophers." The point is that in order to place what is *in principle knowable* to me against the background of what is *in principle unknowable*, I need to see the phenomenon of knowledge somewhat differently. This means nothing less than the recognition of my epistemic position in the sense portrayed above. I am not saying that one has to have a degree in philosophy to venture into this specific territory. I am not inclined to smuggle in the institutional criterion. Yet it must be crystal clear to the person answering the question "what can we know?" that the type of answer one is expected to give should adopt a specific attitude, even if not thought through. It must not be the *natural* attitude like the one referring to documents, or the one referring to banks, job promotions or to who is driving what car, and so on. In other words, all these examples are supposed to *illustrate* something but not to serve as the proper target of problematizations (more on this below).

If the question "what can we know" is apprehended in the way just presented, a provisional understanding of what *knowledge* is (or should be) must be assumed upfront. Otherwise, how are we supposed to draw a line between what is knowable and what is not? However, another possible reading of the question "what can we know?" is something like this: "bring to my attention/understanding what knowledge is." The latter desideratum explicitly spells out a general and abstract puzzle, as if awaiting a more sophisticated intellectual treatment. Yet, again, there is nothing particularly philosophical in this desideratum if we don't take the "what" part in a specific way. For example, this kind of desideratum is likely to drive research in cognitive psychology where specific mechanisms underlying our representations of the world are investigated, thus not so much what knowledge *is* but what the cognitive *mechanisms* that produce it are. Meanwhile, as Fine (2012) put it, philosophers are seriously concerned with *what* questions (what a given entity is) and not *how* questions (how the entity at stake works or how it is constructed). Therefore, taking the "what" phrase seriously requires some sort of extra attention and a decision as to the level and kind of generality that we want to adopt: are we talking, again, about knowledge of banks and cars, knowledge as a practical means to an end, or about "knowledge" referring to the entire domain of what can be known? Are we problematizing a selected part of the cave, to bring Plato in again, and asking what it means to *know that part* as opposed to knowing or not knowing some other parts, or are we problematizing the cave as a whole and asking for an understanding of knowledge that would cover all there is to know? Once more, while the former kind of knowledge operates within the bounds of natural-kind generality, leading to internal questions in Carnap's (1950) sense, the latter one requires another type of reflection, the "second voyage."

Note that the two philosophical concerns, both of which can be articulated by the simple question "what can we know?" maintain a loop connection so that we cannot have one without the other. As mentioned, we need to assume some understanding of knowledge (referring to entities from the inside of Plato's cave) in order to question knowability as such (to question the limits of the cave). On the other hand, we need to have some sort of presumption as to what is supposed to be knowable (a provisional grasp of the limits) if we are to question what knowledge is (describe the type of knowledge that can cover the whole cave). There is no vicious circle, but instead the co-constitution of these two extremely abstract puzzles. This has a certain bearing on Gettier's examples, as we shall see in a moment.

How about Harrell's (2016) other questions? Take "what is real?" If it is not supposed to refer to a real dog taking a walk in the park as opposed to the hallucinatory dinosaur seemingly chasing the dog, we must be able to leap again to the type of generality which would enable us to recognize the whole world as we cognize it *as a problem domain* appropriate for *a real vs. not real* distinction (where "not real" means, for example, something composed solely of sense-data).

Not to spill more ink on that, I provisionally assume that similar lines of argumentation can be assigned to other questions from Harrell's list. We should already have a provisional grasp of the specificity of philosophizing when compared to other intellectual endeavors. In other words, we begin to see the field of philosophy within a broader land cognitively cultivated by humans, where the pursuit of philosophizing takes place. That said, it is still a rather sketchy description. A more "technical" rendering shall be given first in this chapter, with the use of familiar terms—concepts and conceptual frameworks—and then in the next chapter in the less familiar terms of embodied action and sense-making. For even if the proposed account of what philosophers do is roughly correct, it does *not* follow that what they do actually makes any sense. After all, at least three of the listed sources of inspiration, namely Russell, Carnap, and Wittgenstein, were very outspoken critics of their own discipline. So we still need to do more to justify the effort of philosophizing.

4.7 Thinking from Within

Carnap develops an argument against traditional metaphysics that can be called the argument from epistemic limits. The point he makes is that there is something we can sensibly say as we get to know the world better, and there are things we cannot sensibly say. Some parts of philosophy, especially metaphysics, are criticized based on the notion that they attempt to transcend these limits. Wittgenstein's view, with all its specificity, also belongs to this camp, alongside Machery (2017), whom I shall discuss below. Let us stay with Carnap for a moment, though, since answering his points will serve as sort of a template for answering all the arguments from epistemic limits.

I wish to emphasize the following problem. Let A, B, C… refer to fixed conceptual frameworks. Carnap assumes that for any given question, it is posed either in A, or in B, or in C, and so on, or it is posed outside of any framework and is a meaningless utterance. However, for Carnap the boundary between A and B, B and C, and, most importantly, the boundary between the sum of A, B, C, and so on, on the one hand, and the uncountable sum of all meaningless expressions, on the other, seems to be *fixed* and thus unproblematic. In other words, for Carnap the limit of the domain of rational thought, of what can be said, as Wittgenstein (1922) would put it, is given, as if he could see both domains—the meaningful and the meaningless—from an external point of view. But are we in a position to obtain such a glimpse of our conceptual frameworks? Given the very epistemic limits that Carnap himself seems to have in mind, the answer is clear—we are not.

A similar approach to Carnap's distinction was taken by Putnam. He writes that to "distinguish a sense that 'Chairs exist' or 'Material objects exist' *has within the language*, an 'internal' sense, from a pseudo-sense which can only be characterized from outside the language, is precisely to miss the point that there is no standpoint available to us outside the language" (Putnam 2008, p. 8). Following this observation, we need to admit that if all of the subject's intellectual activities are trapped inside conceptual frameworks, then an overview *of these very frameworks* is trapped as well. Thus, generally speaking, an overview of what can and what cannot be achieved by a given conceptual framework is not merely given, but must be actively staked out *from the inside*, so to speak. This is the critical point.

So if we have the conceptual frameworks A, B, C… at our disposal, there is no position outside all of them taken *en bloc*, thus outside the space of what can be said, which would enable us to see them all as if from a God's eye vantage point and to learn about their ties and limits. For this would mean that there is some, as Allison (2006) could say, "theocentric" framework, say G, encompassing A, B, C and so on and at the same time reaching beyond all of them. Even if the very idea of G made sense, contrary to Allison's strong recommendations, we clearly wouldn't be in possession of it.

It merits separate debate as to whether or not, or to what extent, Carnap smuggled elements of the "theocentric epistemology" into his conception. Clearly, he did not say explicitly anything that would suggest he intended to do so (actually, there are reasons to believe that overall he was in the opposite camp). On the other hand, however, the way he speaks of the internal-external boundary and especially his conception of *choosing a language* do make such a suggestion. For if one is supposed to pick a language, it is as if the chosen language was a tool such as a hammer in one's hand, where there is a clear line between the tool, the user, and the external entity the tool is supposed to aim at. Meanwhile, if everything we cognitively achieve is constrained by language, alongside other cognitive constraints, we never just pick a single linguistic tool, which is clearly distinct from the thing we want to grab by it; instead, we wander through various dynamic conceptual realms as if through a thick forest where each entity can be either a tool or something we want to deal with by means of a tool. There is no vantage point we can climb up to in order to see the realm from above. If there is any structure in it, if there are better and worse paths through it, *we must figure them out from within*, metaphorically speaking.

The forest metaphor seems even more apt than Plato's cave, since given what we know about the evolution of cognitive capacities in general, and the history of human epistemic efforts in particular (including the history of science), there are good reasons to believe that the whole structure of cognitive constraints that we are in, including conceptual frameworks, is dynamic, evolving (which is especially stressed by the evolutionary epistemology championed by Lorenz 1977 and Popper 1972, 1984, among others), thus continuously changing (perhaps at times in revolutionary ways, if Kuhn 1962 was right). In effect, there may be no ready-made, perfectly stable order in the domain of what can be said—no pre-given structure or setup of walls which likely characterizes the Platonic cave. Therefore, we shouldn't expect any pre-given instructions as to how we are supposed to navigate the dynamically changing realm constituted by our cognitive tools, including our conceptual frameworks. This is clearly articulated in Wittgenstein's "Philosophical Investigations" (1953), a major difference when compared to the "Tractatus" being exactly his departure from the notion of stable, fixed structures of language, thus also boundaries or limits.

Therefore, going back to Carnap's posit, if we are to figure out which question is internal or external with respect to which conceptual framework, we must search for all of these factors *from within*. Arguably, we therefore need an intellectual pursuit specializing in this specific task.

4.8 Floating Categories

What do we actually mean when we speak of seeking the boundaries or limits of a conceptual framework; when we speak of structures or orders in the conceptual realm? What are we actually referring to when invoking these figurative metaphors? Philosophers have a well-established answer—namely that these structures are constituted by a system of *categories*. The latter can be conceived, following Aristotle, as the most general *de re* "inventory" of all types of beings, thus the sweeping "inventory" of what there is and how it is. Another option is to take them as the most general "inventory" of our ways of speaking, thinking, or cognizing things (thus *de dicto*). In yet another, more functional manner, we can think of categories as providing the most fundamental "architecture" of the edifice of knowledge: they do not fill any of the rooms inside the building, for this is reserved for the empirical disciplines, but they set up the walls separating the rooms, thus creating an organized space. This way or another, the main feature of a system of categories is that either reality itself or our intellectual effort to capture it (or both) is organized, and this organization can be unpacked by means of specific concepts. This means that there is some order spelled out in terms of not only natural-kind concepts such as CHAIR, TREE, or TIGER, but also a more fundamental "plan" organizing these concepts and/or their referents. Aristotle distinguishes ten categories, including, inter alia, SUBSTANCE, QUANTITY, QUALITY, RELATION, STATE, or PLACE. This means that for any entity *x* that can be assigned a description, we can ask, for example, *what* it is (an individual substance, a state, a property, or relation) and *where* it is. Sitting at my desk is a state located in a specific room, but perhaps it may also be taken as a relation between me and my desk. So the idea is that while we all speak of numerous kinds of relations, and while scientists investigate some of them, the remaining question is what is *a relation as such*, independently of all its instantiations. Similarly, we all talk of different things, but what is *a thing*—a substance—*as such*? Provided that such questions are legitimate—which is a controversial subject, as we saw already when citing Carnap and others—they refer to categories.

Newer systems of categories include the proposals of Brentano (1981), Husserl (1901/2000), Hartmann (1940), Ingarden (1964/2013), Chisholm (1996), or Lowe (2006), among others. For example, Jonathan Lowe distinguishes PARTICULARS and UNIVERSALS in the first place; then in the first group OBJECTS and MODES, and KINDS alongside ATTRIBUTES in the second group.

However, categories themselves are not our major concern. Philosophers working with them have to answer various challenges I find marginal from the standpoint of this investigation. For example, a system of categories is traditionally expected to be complete and, crucially, justified in such a way that no doubts arise as to its validity. The latter is another challenge—what method of reasoning or insight can we use to come up with an ultimate inventory immune to any future empirically based correction? Not without merit, some scholars have serious doubts as to whether the pursuit in question can be conclusive. Meanwhile, all these ambitions are abandoned in my project, primarily due to what I take as fact—flagged a moment ago—namely that our conceptual frameworks, and more generally all our cognitive constructs, are dynamic, continuously changing, so that we can hardly speak of any clear-cut conceptual order, allegedly established once and for all. All we have are *relatively* stable structures.

Instead, I propose a more modest approach, which at the same time leaves quite a lot of room for traditional philosophizing. To begin, I assume that we can distinguish within every fixed group of conceptual frameworks, at a particular stage of their development, a group of *boundary concepts* for these frameworks. The phrase "for" is to indicate that they perform a certain *job* for the conceptual structures; they are not merely there, they are there for a purpose.

The notion of boundary concept has not attracted the attention it deserves from philosophers, I believe. A path to a proper philosophical analysis has been set forth by Sebastian T. Kołodziejczyk (2006) who is especially interested in what traditional metaphysics calls transcendentals. I draw inspiration from his work, however, I shall take a more pragmatic stance here.

Here is a nice characterization of boundary concepts formulated by researchers working in another field, far from philosophy, a field in which the *job-related* aspect mentioned above certainly proves critical, that is, by two researchers working in the inherently multidisciplinary field of conservation biology and conservation policy:

> A boundary concept enables researchers from different disciplines, policy makers, and other stakeholders to develop a common language, and respectively integrate and derive knowledge relevant to their field. At the same time, the exact meaning and conceptualisation of such a boundary concept is contested both between different scientific disciplines, and between science and society; this will be determined in boundary negotiations. (Schleyer et al., 2017, p. 3; online version)

Note the pragmatic stance manifested in these words. There is no pre-given, ready-made meaning of a boundary concept which the users of the two (or more) conceptual frameworks are merely expected to master. The meaning is something that *happens*, so to speak, in-between the cognitive pursuits of both groups. I shall uphold this pragmatic stance.

It would be good now to turn to an example. Consider the conceptual framework of microeconomics, composed of such notions as SUPPLY, DEMAND, PREFERENCE, COST, and PROFIT. Each of these notions is supposed to capture phenomena observed on the market. But what about the concept of MARKET itself? Note that, perhaps somewhat surprisingly, it does *not* belong to microeconomics in the way SUPPLY or PROFIT do, for it is not supposed to capture any *specific* events or processes investigated by microeconomics; thus it is not defined there. Instead, a proper understanding of the concept MARKET is necessary for a proper understanding of each of the microeconomic concepts listed above and for a proper understanding of what microeconomics is about in the first place. In other words, the concept MARKET does not *refer* to any specific entity or phenomenon targeted by microeconomics, yet all the entities and phenomena targeted by microeconomics *take place on* the market. To put it differently still, if conceptual frameworks are represented as clusters of meaning rules, in step with Ajdukiewicz (1934), there is no separate meaning rule within microeconomics defining how the term "market" is to be used properly, while at the same time its proper use is a necessary condition of all specific rules referring to terms like PROFIT or SUPPLY. Actually, there is a variety of such notions codetermined by one another, for example, EXCHANGE or GOOD. What is a market? It a place where goods are exchanged by buyers and sellers. What is a good—it is an entity of any kind that is exchanged on the market. What is an exchange? It is a process in which a good is transferred from a seller to a buyer on the market. Therefore, the concept MARKET is defined contextually within a group of similarly abstract notions, all of which are operationalized in various ways by microeconomics, without any of them actually being independently defined there.

Now, while asking a question like "is there a market for private health services in Poland?" does make sense in accordance with Carnap's standards, that is, it is an internal (internal-to-microeconomics) question, the puzzle "what is a market?" is not internal in the same sense. Rather, it is a boundary question, since it targets a *boundary* concept.

Having said all this, the somewhat non-orthodox nature of the proposal set forth here manifests itself in the postulate that we should *not* expect any system of categories thought of as a system of boundary concepts within fixed dynamic conceptual frameworks to be either complete or everlasting. Depending on the stage of the relevant conceptual dynamics, various concepts can *become* boundary concepts, and various boundary concepts can turn out to be general and abstract enough to justify a relatively independent (philosophical) intellectual effort. To illustrate this let us consider another example: INFORMATION. While its career begins from the strictly technical account by Shannon and Weaver (1949), these days it is hard to find a phenomenon that is not targeted in terms of information, beginning with quantum information, through genetic information, up to semantic information, information-processing representational systems, the information disparity crippling economic decisions, and information cocoons created by social media (see Floridi 2011). And this is not even a remotely complete list. Some people even go as far as to argue that information may have to do with the ultimate metaphysical level of reality and God (see Davies and Gregersen 2010). This I take as an indication of the fact that INFORMATION has become a boundary concept playing the categorial role at the present moment in history.

Someone might ask, however, what does this have to do with philosophy? If we think of, for example, the concept MARKET, which does not belong to just one specific conceptual framework, there is, indeed, a need for some specialized boundary-targeting intellectual pursuit *within* a variety of disciplines (economics, sociology, etc.), but still no need for philosophy as a separate field. The same applies, for example, to the concept of information—someone might argue that Shannon and Weaver's (1949) theory laid the groundwork for many future developments of the concept within specific fields and that there is no need for philosophy.

A minimal answer to these points would be that there is a method or perhaps a heuristic used in all or in a large number of such boundary-targeting analyses across various disciplines, and that it is of some practical value to have people skilled in it. These people are conventionally called *philosophers*. Therefore, we end up with a picture of philosophizing as a somewhat dissolved cluster of pursuits called "philosophy of economics," or "philosophy of biology" (targeting such boundary concepts as ORGANISM, BIOLOGICAL INDIVIDUAL or ENVIRONMENT), "philosophy of physics" (concerned with CAUSATION, TIME, SPACE), and so on. They all maintain some sort of family resemblance in the sense of Wittgenstein (1953), which in turn gives rise to a certain division of labor and institutional regimes in academia. And that's basically all there is to say about the role of philosophizing.

This minimal answer is not wrong, I believe, yet I find it insufficient. For we need to dig deeper to understand what constitutes the said family resemblance, provided that the term "philosophy" is not merely shorthand for a few dozens of words that would be more fine-grained. I argue that there *must be* something deeper underneath the family resemblance, something that makes all these different philosophical endeavors, dispatched, as it were, across various scientific disciplines *possible*; a factor such that, should it be missing, none of these dispersed pursuits could have emerged. This factor is already familiar to the reader, at least to some extent: it is the recognition of one's epistemic position.

I substantiate the latter point in an indirect way. Namely, I argue against the position called meta-philosophical minimalism recently put forward by Edouard Machery (2017). Machery argues that there is no philosophy-specific act of intellect or attitude whatsoever. Thinking philosophically is just like any other kind of thinking, and thus the difference between philosophy and other disciplines refers to the topics philosophers are interested in. Although it sounds benign, this claim goes directly against the idea rolled out above, and therefore calls for an answer. For while there is a clear difference in the topics or problems (no doubt about it), first, these problems did not fall from the sky—they had to be brought forth by some intellectual effort and the core debate in meta-philosophy is actually about the latter. Second—we have already observed that these topics or problems are either apparent trivialities, for example, "what is real?", or can equally serve as targets of other disciplines, for example, "what is knowledge?" So if Machery—and Russell (cited earlier)—is right that there is no specifically philosophical pursuit of the intellect, one that uniquely produces philosophical problems, there is no substance in postulating philosophy as a relatively independent discipline of thinking. Instead, it should be dispatched in the way illustrated above, adding some conceptual self-awareness to each specific discipline. Therefore, arguing successfully against Machery's meta-philosophical minimalism would at least open up a prospect for substantiating the opposite claim to the effect that there is a need for a relatively self-standing discipline of philosophy.

4.9 Machery on the Method of Cases: The Problem of Comparability Reconsidered

Machery begins by analyzing the *method of cases* which is, in his view, a common strategy of argumentation employed by philosophers (in particular twentieth-century Anglo-Saxon analytic philosophers, as Machery admits). Provided that Machery is right on this point, we can understand the call issued by Xphi. This call can be seen as a challenge to the apriorism of traditional philosophers, arguing that the cases studied by them should be empirically tested. If philosophy is indeed built on an analysis of specific cases, then Xphi's call seems, indeed, reasonable and appealing.

Let us then begin by asking what these philosophical cases are, thus, indirectly, what the topics that carve up philosophizing from a wider set of intellectual endeavors are. The problem is that here Machery, in my reading, adopts a position that may be deemed meta-philosophical apriorism, which in and of itself is surprising given his commitment to Xphi. For when he asks, first, which cases are *philosophical*, the prompt answer is that "philosophical cases are cases put forward by philosophers" (Machery 2017, p. 11). The meaning of "philosophers" seems to be set, having, most likely, the institutional sense sketched out in Chaps. 1 and 2. So the assortment of problems or topics deemed philosophical is constructed purely a priori—it is, admittedly, grounded in well-established academic practice and literature, yet independent of particular subjects' engagement in reflection; independent of their actual desiderata and experiences, as well as the circumstances in which a given act of reflection takes place. Certain topics just *are* attributable to academic philosophy, full stop, while the way they were brought forth in the first place is irrelevant. This is somewhat contrary to the spirit of Xphi which began as an attempt to leave the symbolic armchair and check out how people *actually think*. Why does the latter ("how people think") include only people's answers to ready-made questions, but not how *they come to ask* those questions in the first place? It can always be said that the relevant questions are after all delivered to the participants of Xphi studies by vignettes, but this cannot be a good reply—if one is supposed to give an answer, one actually has to *ask the question* from the vignette, thus to come up with a specific kind of desideratum.

Going back to what Chap. 3 puts forward, especially based on Rescher (2000), desiderata are requests for information. Their general form is something like this: "bring it about that I know a fixed condition C_1 *as contrasted with* a (fully or partially specified) set of alternative conditions $C_2,\ldots,C_n$." Being able to come up with a desideratum is due to a very specific cognitive faculty dubbed erotetic intuition in Chap. 3. All that being said, if one is supposed answer a question referring to a particular philosophical case, one needs to kick-start one's erotetic intuition resulting in the formulation of a specific kind of desideratum first.

Yet we learn nothing about the formulation of a desideratum from Xphi studies, as if they assumed a priori that the desiderata that folks come up with are exactly those they are expected to come up with (see Chaps. 2 and 3). A proponent of Xphi might protest at this point and rush to clarify that there is no expected desideratum, yet this reply is also a non-starter. For if there is no expected desideratum, people are free to associate the vignette queries with anything they fancy, which would put their replies very much in peril of what expertise defenders think of them. Let us keep these concerns in mind for a while as we proceed.

Aside from what makes a case philosophical, what is *a case* as such? It is a short description of a hypothetical situation, including one that does not incite philosophical reflection. Now, as for the characterization of *philosophical* cases, according to Machery, they boil down to descriptions of hypothetical situations, sometimes very unlikely or even at odds with the laws of nature, whose purpose is to highlight *necessary conditions* of this or that phenomenon (e.g., what are the necessary conditions of having knowledge). So despite being initially rather enigmatic as to what the qualification "philosophical" is supposed to stand for, here Machery is more specific—what makes a case philosophical is the pretense to unveil modal qualities of real phenomena, for example, what consciousness is *necessarily* like and what knowledge *necessarily* is.

That said, Machery deems the traditional method of cases "modally immodest" precisely due to this pretense to modal knowledge, since it aims at something that, as he says, is "beyond our epistemic reach." Thereby Machery articulates yet another version of the argument from epistemic limits, following the path established by Kant, Wittgenstein, and Carnap. As before, here, too, it seems to be assumed upfront what these limits are, as if they were fixed once and for all, which is quite problematic, but I shall not reiterate the points already spelled out earlier. What is significant here is the puzzle of whether or not any *specific* intellectual effort, different from other intellectual efforts people make, is required in order to properly apprehend philosophical cases. Let me stress though that speaking of a "specific intellectual effort" does not mean a "better" intellectual effort, somehow transcending the limits of our epistemic position. If there is any philosophy-specific intellectual attitude, it is as "good" as any other (whatever that might mean); the debate being over its specificity, not over its alleged over-the-top pretenses.

So, the question is this: "What kind of state is the reader supposed to have about the situation described by a philosophical case?" (Machery 2017, p. 12). Machery proceeds to answer. First, the reader can "reason about this situation, draw inferences about it, and categorize its elements," (ibid.) and while doing so, the reader is "using the very processes we use toward situations we believe to be actual" (ibid.). Still, even if that's the case, someone might argue that there is some specific attitude attributable to these otherwise quite usual procedures, that is, reasoning or concept application. Aware of this option, Machery goes on to spell out and advocate the minimalist approach, as he dubs it, according to which "philosophical cases do not elicit attitudes distinct in kind from the application of concepts in everyday life" (ibid., p. 20). At the other extreme, the exceptionalist approach argues that "the attitudes elicited by philosophical cases differ from everyday judgments, i.e., the kind of judgment we routinely make when we apply concepts to possible instances" (ibid., p. 17).

Now, there are, in Machery's view, three ways in which philosophical judgments can be considered exceptional: either they have a specific phenomenology, a sort of feel or sense of necessity attributable to them, or a distinct epistemic role, such that, for example, unlike other judgments they are justified purely a priori; or they may have a distinct etiology by expressing exceptional intuitive competence regarding a given concept, for example, knowledge. He then argues, convincingly, that neither of these exceptionalities actually applies to philosophical judgments. No such phenomenology, no such a priori intuition and no such exceptional competence have ever been proven empirically to exist, he insists. Fair enough, I think that Machery is correct here.

All that being said, what I would like to do here is not to defend metaphilosophical exceptionalism, but to demonstrate that the recognition of one's epistemic position—no matter whether exceptional or not, widespread or strictly limited to professional philosophers—is a necessary condition of a proper act of questioning to which a philosophical judgment responds. This is not supposed to invest philosophy with some mysterious, extramundane status, but to show its distinctness. Eventually, the goal of this reflection is to substantiate the point that philosophy as a relatively independent discipline is required if we are to be able to properly articulate boundary questions—questions aiming at the most general architecture(s) of the world as it manifests itself to us through our cognitive framework(s).

The issue of comparability returns at this point, for there is no guarantee that two propositions that look the same *when uttered* are indeed comparable. It all depends on their desiderata. That said, the question is whether or not the condition of comparability is satisfied when it comes to the philosophical cases examined by Xphi, upon which Machery builds his argumentation. As I said, my purpose is not to argue that the condition is not satisfied, for perhaps it is in many of the cases studied by Xphi. The point is that it *might* also *not* be satisfied, as I demonstrated in Chap. 2. So the proper puzzle is what makes judgments or beliefs referring to hypothetical cases comparable when they are, and what differs between them when they aren't.

4.10 Philosophical Cases and Two Kinds of Desiderata

To begin with, judgments such as "Smith knows who gets the job" or "Smith does not know who gets the job," referring to the character from one of Gettier's scenarios, seem to be answers to the same simple question, namely "Does Smith know who gets the job?" But the point is, as I argue below, that the latter can articulate at least two different desiderata that Smith *himself* could have in mind; and in addition, there are different desiderata that *we* may have in mind when *we*, the readers, make judgments about the story. Moreover, provided that this plurality of desiderata is possible, the list of what we, the readers, can intellectually do with a philosophical case needs to be extended. It cannot boil down to reasoning about the case or concept application, which is supposed to lead up to certain judgments, for if these judgments are supposed to be comparable, there should be a cognitive process that makes *an assortment of desiderata*.

I shall now explain how the mentioned plurality of desiderata is possible. First, let us remind ourselves that with respect to any fictional story, we can draw a distinction, elaborated on by Ingarden in his classic "The Literary Work of Art" (Ingarden 1931/1973), between the truth (or falsity) attributable *to* a given story as such, and the truth attributable to sentences *within* the story (for more on truth in fiction see first of all David Lewis' 1978 work; also Hanley 2004; Woodward 2011). This shall be a template for me for making a similar distinction in a moment. For example, we assume that the sentences in C. S. Lewis' "The Chronicles of Narnia" are not true in the real world; that is, it was not the case that a group of kids really went to Narnia via an old wardrobe, yet we do find true propositions *within the story*. So when Lucy says she saw a lion, her statement is true, while the others wrongly take it as false, and all these epistemic attributions are perfectly sensible *within the story*. Now, a similar difference applies to questions and desiderata—we need to distinguish a desideratum *expressed by us*, the readers, *with respect to the story* and the characters populating the story, and *a* desideratum that the characters themselves can articulate *within the story*. The latter in practice means that we might formulate this kind of desideratum if we placed ourselves as a fictional character within the story. Given that difference, the issue of the assortment of comparable desiderata becomes even more pressing, for in addition to these desiderata being simply different in their content, they may be also different in their more fundamental character, namely their "location."

Once we have these two kinds of desiderata, we must also distinguish between the two broader contexts to which they belong: the epistemic position *of* a fictional character *as recognized by us* and the epistemic position of that fictional character as exploited (not necessarily recognized at all) *by the character* in question.

Therefore, I argue that the reader of Gettier's story about Smith pondering who gets the job can do two things: generally speaking, they can either place themselves within the story, or not. The way Gettier sets up his cases suggests the latter, but based on what Nichols et al. (2003), for example, write, the participants are not explicitly instructed not to sympathize more with the character (Smith, in this case). What if they do sympathize with him? Smith's desideratum is something like "Bring me to know who gets the job" and the question asked by the experimenters, that is, "Does Smith know who gets the job" can be understood as "Is Smith's desideratum satisfied?" Now, the answer to that depends on what Smith's desideratum actually is, for it could be something like "Bring me to know whether a person with five dollars in their pocket will get the job," which indicates that—for whatever reason—he *is actually interested in* the money criterion, thus with the satisfaction of a particular predicate/description, not with the name of the lucky individual who will be hired. In this case, the predicate is satisfied, thus I would say that Smith knew what he wanted to know. But a more real-life scenario would be when the desideratum debunks Smith's real interest, namely "Bring me to know whether I get the job or Jones gets the job," in which case I would say that the desideratum is not satisfied—Smith has overplayed his "logical" juggle with predicates and missed the person who actually gets the job (namely, himself).

Let me stress the "I would say" part above—the intention here is not to make any general claim without empirical evidence, as if I had some kind of mysterious access to people's minds. Yet a first-person analysis is sufficient to show that it is possible to obtain two opposite answers simply by assuming two different desiderata that can stand behind Smith's epistemic request for information—or actually behind ours *once we sympathize with Smith* in one way or another and become involved in the story. Meanwhile, Xphi has no tool for staking out who is having which desideratum in their minds.

The real problem is that both of these desiderata are most likely *not* what philosophers have had in mind when reflecting on Smith's situation. For *they are not sympathizers*; Smith's position is not one to place oneself in, but to recognize something *about* and to be used to illustrate a more general point. This, however, is a completely different erotetic situation.

So, in addition to Smith's own desiderata within the story, there are desiderata that are *external to the story*—likely many such desiderata, although here we are interested only in those that epistemologists may have in mind. Speaking of the latter, as indicated in 4.6, questioning knowledge—which is what Gettier's stories are for—may refer to at least two correlated, yet different puzzles: firstly, the issue of the *limits* of knowledge (what is *knowable*), in which case some provisional understanding of what knowledge is has to be assumed; or, secondly, the very concept of knowledge, in which case some provisional grasp of what this concept covers needs to be assumed. Crucially, in both cases we must recognize Smith's epistemic position—we must ask ourselves what was *in principle* within *his* epistemic reach, for example, whether or not it was possible for him to learn who had what amount of money in their pocket, about his boss' intentions, and so on. Remember, however, that this is not supposed to be a recognition made *by* Smith himself or by us sympathizing with Smith; it is made by us qua *external*, God's eye view spectators. Note an interesting maneuver here: while we are unable to actually occupy a God's eye vantage point that would enable us to take a view-from-nowhere glimpse at *our* real epistemic position, we fancy a simplified scenario of Smith with respect to which we can play the role of God-like spectators, using this simplified story to problematize some aspects of our real epistemic position. Arguably, this is the very goal of philosophical thought experiments.

Provided that Smith's epistemic position is clear *to us* (not necessarily to him), we can come up with two external-to-the-story desiderata. First: "Bring to my understanding whether or not the concept of knowledge that we apply to Smith's case is sufficient to cover everything that was within his epistemic reach." The consensus among epistemologists in the aftermath of Gettier's paper was rather that the classic approach is not sufficient: this means that once *we* recognize *Smith*'s epistemic position we can see that the concept of knowledge we intend to apply to his case does not cover what was in principle knowable to him. For example, Smith could have done more to check out who satisfied the description "a person with five dollars in their pocket." Alternatively, Goldman (1979) have proposed that a new condition of reliability needs to added, which is a major change, for it extends the limits of what was in principle knowable to Smith beyond the limits of his awareness. Let me reiterate: the issue here is to come up with a characterization of knowledge that would cover everything within Smith's epistemic reach that *we* can recognize from our imaginary God-like vantage point.

Note however, that Smith's case can also be interpreted the other way around (as described in 4.6): provided that the classic concept is set, the *range* of knowability may be problematized. So, secondly we can have a desideratum like this: "Bring me to understand what was within Smith's epistemic reach, thus what he could know." It is not about what he did to justify his claims, nor about the concept of knowledge. The concept is assumed, and "Does Smith know?" means "is this or that thing within Smith's epistemic reach, given such and such an understanding of knowledge?" If we say "He does not know who gets the job" in this context, we are not making it an issue whether our definition of knowledge is correct, but whether some facts were epistemically available to Smith. That said, own concern was clearly about the concept of knowledge.

In sum, the whole point of philosophically examining stories such as those written by Gettier is *for us* to recognize *the character's* epistemic position, yet the recognition is *not* part of the story. Alternatively, we could sympathize with the character, place ourselves in their position, which is a different attitude or a different stance. While the latter gives rise to desiderata that are internal to the story, the former brings forth desiderata that are external to the story. In a sense, the distinction mirrors

Carnap's famous distinction elaborated on earlier. This, I propose, is the core difference between the philosophical attitude and non-philosophical attitudes—the difference denied by Machery (2017). The difference is not in some mysterious feeling of necessity or any other extramundane epistemic capacity that perhaps some philosophers attribute to themselves. It is much more down-to-earth. Let me reiterate: it's about different desiderata. Noteworthily, there is nothing in it that would constitute an *in principle* problem to Xphi. As long as the experimenters know and somehow control the desiderata formulated by their subjects, Xphi can without doubt contribute to epistemological debates. As I have indicated several times, this investigation is neither pro nor contra Xphi, but rather sets a new background against which the status of philosophizing needs to be re-recognized or restated.

What does all of this have to do with the things raised in the previous section, thus with the issue of boundary questions and the need for a relatively independent discipline devoted to them? The answer is quite straightforward: we can now see that the recognition of one's epistemic position plays a foundational role if we are supposed to come up with a boundary question. In Gettier's stories it was the recognition of their characters' epistemic positions, yet what it really was about was *our* epistemic position. KNOWLEDGE is a boundary concept unveiling, alongside such notions as TRUTH or JUSTIFICATION (or, e.g., the notion of reliable causal process, if we follow Goldman), the most fundamental "architecture" of our cognitive efforts, of the realm covered by our cognitive acts. As trivial as it may sound, what epistemologists are concerned with is not Smith, Jones, or any other character populating their illustrative stories, but the question of what notions link and uphold, thus make possible, the entire discourse referring to human (and possibly not only human) cognitive endeavors. Therefore, the boundary question of knowledge, truth, and so on never just drops from the sky, nor is it found lying on the street ready to be picked up. There needs to be an intellectual effort put into problematizing the very fact that we *are* cognizers; an intellectual effort whose distinctness is not (at least not necessarily) grounded in any pretense to a qualitatively superior insight, but in the recognition of the philosophizing subject's epistemic position as manifested in specific desiderata. That said, there is no reason to deny the possibility that people who are not professional philosophers could do this specific thing, nor is there reason to optimistically assume that all professional philosophers actually do it. The point is that it should somehow be controlled as part of both experimental and armchair investigations.

What applies to the boundary concept of knowledge applies to virtually all concepts analyzed by philosophers. This includes more specific notions attributable to dispersed applied philosophies discussed earlier, for example, such concepts as INFORMATION, as it stands in various disciplines, ORGANISM as used by biology. They all share the same quality *qua boundary concepts* whose problematization requires peculiar intellectual effort. And for that reason, philosophy is needed, I argue, as a relatively independent discipline, operating at the level and with the kind of generality that can only arise in the recognition of one's epistemic position. As such, philosophy *per se* does not boil down to these applied philosophies, but instead makes the latter possible. More on this in the next chapter.

4.11 Conclusion: Erotetic Intuition Out of Our Heads

This chapter was intended to convey one key idea: what it means for a subject to philosophize is not fixed primarily by any a priori institutional division of labor and assortment of ready-made topics, but is grounded in a peculiar act of that subject's erotetic intuition—the recognition of his or her epistemic position. What that means, in short, is becoming aware of the fact that the world in which the subject lives, the target of all her behaviors and cognitive acts, is a certain presentation of an assumed reality, determined by the subject's location and cognitive capacities. This is the discovery made by the cave-dwellers from Plato's metaphor: the recognition that they, indeed, *are in a cave*. Yet, unlike Plato, whose optimism in this respect also decreased over time, we have no reason to believe that leaving the cave is even a thinkable option. Therefore, in addition to specific, well-defined, and well-focused disciplines devoted to orienting the cave-dwellers inside their environment, we need a discipline whose very purpose is to uphold the awareness that the cave *is a cave*, however trivial that may sound. For otherwise the cave-dwellers would have been unable to elicit and problematize the fact that the pursuit of orientation within the cave is made possible by more fundamentals factors attributable to the cave as a whole.

This characterization of the pursuit of philosophizing is still rather enigmatic, however. We should be able to say more, to "operationalize," as it were, the notion of recognition of one's epistemic position. I see at least two options here. One would be a more technically developed theory of philosophical questions, grounded in and making use of some of the already developed fragments of erotetic (epistemic) logic. This is the path taken by Rescher (2000) or Perzanowski (1990, 2012).

There is a limit to this logical approach, however, despite the fact that the need for it is undeniable in light of Carnap's and Wittgenstein's challenges. Namely, it leaves the most troubling issue discussed so far in this book untouched; it has no tools to capture the very cognitive capacity of asking questions; of problematizing selected parts and aspects of our surroundings. Thus is leaves untouched the cognitive faculty provisionally called erotetic intuition.

Therefore, the next chapter takes a different path. It focuses not on questions as products of a certain cognitive capacity or of certain cognitive acts, but precisely on these very acts. We have already put forward the idea that actual desiderata, including those that philosophers come up with, are somehow embedded in people's actual, situation-dependent epistemic needs. So the somewhat mysterious capacity of erotetic intuition shall be accounted for—at its core—as a way of dealing with specific circumstances. This means that it is not primarily a mechanism in our heads, to put it figuratively, but a mechanism embodied in action, constituted by the cognizer's embeddedness in specific environments. We shall seek a concept of the embodied philosopher; that is to say—the philosopher approached as a living creature coupled with its surroundings.

References

Ajdukiewicz, K. (1934). Das Weltbild und die Begriffsapparatur. *Erkenntnis, 4*(1), 259–287.

Allison, H. E. (2006). Kant's Transcendental Idealism. In G. Bird (Ed.), *A Companion to Kant* (pp. 111–124). Blackwell.

Andow, J. (2016). Qualitative Tools and Experimental Philosophy. *Philosophical Psychology, 29*(8), 1129–1141.

Ben-Zeev, A. (1984). The Passivity Assumption of the Sensation-Perception Distinction. *The British Journal for the Philosophy of Science, 35*(4), 327–343.

Brentano, F. (1981). *The Theory of Categories*. Nijhoff.

Brown, D. (2008). Descartes on True and False Ideas. In J. Broughton & J. Carriero (Eds.), *A Companion do Descartes* (pp. 196–215). Blackwell.

Carnap, R. (1928/1967). *Der Logische Aufbaum der Welt*. Felix Meiner Verlag. Quotations from: R. Carnap, *The Logical Structure of the World. Pseudoproblems in Philosophy*. University of California Press 1967.

Carnap, R. (1950). Empiricism, Semantics, and Ontology. *Revue Internationale de Philosophie, 4*, 20–40.

Chisholm, R. (1996). *A Realistic Theory of Categories*. Cambridge University Press.

Chudnoff, E. (2011). What Intuitions Are Like. *Philosophy and Phenomenological Research, LXXXII*(3), 625–654.

Crane, T. (2001). *Elements of Mind*. Oxford University Press.

Davies, P., & Gregersen, N. H. (Eds.). (2010). *Information and the Nature of Reality: From Physics to Metaphysics*. Cambridge University Press.

Fine, K. (1982). Acts, Events, and Things. In W. Leinfellner, E. Kraemer, & J. Schank (Eds.), *Language and Ontology. Proceedings of 6th International Wittgenstein Symposium* (pp. 97–105). Hölder-Pichler-Tempsky.

Fine, K. (2012). What Is Metaphysics? In E. T. Tahko (Ed.), *Contemporary Aristotelian metaphysics*. Cambridge University Press.

Floridi, L. (2011). *The Philosophy of Information*. Oxford University Press.

Goldman, A. (1979). What Is Justified Belief? In G. S. Pappas (Ed.), *Justification and Knowledge* (pp. 1–25). D. Reidel.

Hanley, R. (2004). As Good As It Gets: Lewis on Truth in Fiction. *Australasian Journal of Philosophy, 82*(1), 112–128.

Harrell, M. (2016). *What Is the Argument. An Introduction to Philosophical Argument and Analysis*. MIT Press.

Hartmann, N. (1940). *Der Aufbau der realen Welt. Grundriss der allgemeinen Kategorienlehre*. De Gruyter.

Horner, C., & Westacott, E. (2000). *Thinking through Philosophy. An Introduction*. Cambridge University Press.
Husserl, E. (1901/2000). *Logische Untersuchungen. Zweiter Teil: Untersuchungen zur Phänomenologie und Theorie der Erkenntnis*. Halle: Max Niemeyer. Available in English *Logical Investigations* (2nd ed., 2 volumes) (J. N. Findlay, Trans.). Routledge.
Husserl, E. (1913/1983). *Ideen zu einer reinen Phänomenologie und phänomenologischen Philosophie, I, Buch: Allgemeine Einführung in die reinen Phänomenologie*. Halle: Max Niemeyer Verlag. Quotations from: E. Husserl, *Ideas Pertaining to a Pure Phenomenology and to a Phenomenological Philosophy. First Book: General Introduction to a Pure Phenomenology*. The Hague/Boston/Lancaster: Martinus Nijhoff Publishers/Kluwer, 1983.
Husserl, E. (1950/1982). *Cartesianische Meditationen*. Den Haag: Martinus Nijhoff. Quotations from: E. Husserl, *Cartesian Meditations. An Introduction to Phenomenology*. The Hague/Boston/London: Martinus Nijhoff Publishers/Kluwer 1982.
Ingarden, R. (1931/1973). D*as literarische Kunstwerk. Eine Untersuchung aus dem Grenzgebiet der Ontologie, Logik und Literaturwissenschaft*. Max Niemeyer, 1931. Available in English: *The Literary Work of Art: An Investigation of the Borderlines of Ontology, Logic, and Theory of Language* (trans by G. G. Grabowicz). Northwestern University Press.
Ingarden, R. (1964/2013). *Der Streit um die Existenz der Welt* (Vol. 1: Existenzialontologie). Tübingen: Max Niemeyer. English translation: *Controversy Over the Existence of the World* (Vol. I). Frankfurt: Peter Lang.
Kołodziejczyk, S. T. (2006). *Granice pojęciowe metafizyki* [The Conceptual Boundaries of Metaphysics]. Uniwersytet Wrocławski.
Kraemer, H. J. (1990). *Plato and the Foundations of Metaphysics. A Work on the Theory of the Principles and Unwritten Doctrines of Plato with a Collection of the Fundamental Documents* (J. R. Catan, Trans.). State University of New York Press.
Kuhn, T. (1962). *The Structure of Scientific Revolutions*. University of Chicago Press.
Lewis, D. (1978). Truth in Fiction. *American Philosophical Quarterly, 15*(1), 37–46.
Lorenz, K. (1977). *Behind the Mirror*. Methuen.
Lowe, E. J. (2006). *The Four-Category Ontology: A Metaphysical Foundation for Natural Science*. Clarendon.

Machery, E. (2016). De-Freuding Implicit Attitudes. In M. Brownstein & J. Saul (Eds.), *Implicit Bias and Philosophy. Volume 1. Metaphysics and Epistemology* (pp. 104–129). Oxford University Press.

Machery, E. (2017). *Philosophy Within Its Proper Bounds*. Oxford University Press.

Merleau-Ponty, M. (1945/1962). *Phénoménologie de la perception*. Available in English: *Phenomenology of Perception* (C. Smith, Trans.). Routledge & Kegan Paul.

Nagel, T. (1986). *The View from Nowhere*. Oxford University Press.

Nichols, S., Stich, S., & Weinberg, J. (2003/2011). Metaskepticism: Meditations in Ethno-Epistemology. In S. Luper (Ed.) *The Skeptics* (pp. 227–247). Ashgate Press. Quotations from S. Stich, *Collected Papers. Vol. 2. Knowledge, Rationality and Morality, 1978–2010*. Oxford: Oxford University Press, 2011.

Osler, M. J. (2008). Descartes' Optics: Light, the Eye, and Visual Perception. In J. Broughton & J. Carriero (Eds.), *A Companion do Descartes*. Blackwell.

Perzanowski, J. (1990). Ontologies and Ontologics. In E. Żarnecka-Biały (Ed.), *Logic Counts* (pp. 23–42). Kluwer.

Perzanowski, J. (2012). *Art of Philosophy. A Selection of Jerzy Perzanowski's Works* (J. Sytnik-Czetwertynski, Ed.). Ontos Verlag.

Popper, K. R. (1972). *Objective Knowledge: An Evolutionary Approach*. Clarendon Press.

Popper, K. R. (1984). Evolutionary Epistemology. In J. W. Pollard (Ed.), *Evolutionary Theory: Paths into the Future* (pp. 239–255). John Wiley & Sons.

Putnam, H. (1987). *The Many Faces of Realism*. Open Court.

Putnam, H. (1990). Realism with a Human Face. In H. Putnam, *Realism with a Human Face* (edited by J. Conant) (pp. 3–29). Harvard University Press.

Putnam, H. (2008). Wittgenstein and Realism. *International Journal of Philosophical Studies, 16*(1), 3–16.

Reale, G. (1990). *A History of Ancient Philosophy*. Vol. II. *Plato and Aristotle* (J. R. Catan, Trans.). State University of New York Press.

Rescher, N. (2000). *Nature and Understanding*. Oxford University Press.

Russell, B. (1912/2004). *The Problems of Philosophy*. Williams and Norgate. Quotations from: B. Russell, *The Problems of Philosophy*. New York: Barnes & Noble, 2004.

Schleyer, C., Lux, A., Mehring, M., & Görg, C. (2017). Ecosystem Services as a Boundary Concept: Arguments from Social Ecology. *Sustainability, 9*, 1107. https://doi.org/10.3390/su9071107

Searle, J. R. (1992). *The Rediscovery of Mind*. MIT Press.

Searle, J. R. (1998). *Mind, Language and Society: Philosophy in the Real World*. Basic Books.

Sellars, W. (1956). Empiricism and the Philosophy of Mind. In H. Feigl & M. Scriven (Eds.), *Minnesota Studies in the Philosophy of Science* (Vol. I, pp. 253–329). University of Minnesota Press.

Shannon, C. E., & Weaver, W. (1949). *The Mathematical Theory of Communication*. University of Illinois Press.

Tarski, A. (1983). *Logic, Semantics, Metamathematics: Papers from 1923 to 1938*. Hackett Publishing.

Twardowski, K. (1911/1999). O czynnościach i wytworach. Kilka uwag z pogranicza psychologii, gramatyki i logiki. Originally published in 1911; translation available in: K. Twardowski, *On actions, products and other topics in philosophy*. Rodopi.

Varela, F., Thompson, E., & Rosch, E. (1991). *The Embodied Mind: Cognitive Science and Human Experience*. MIT Press.

von Uexküll, J. (1934/2010). *A Foray into the Worlds of Animals and Humans with a Theory of Meaning*. University of Minnesota Press.

Wittgenstein, L. (1922). *Tractatus Logico-Philosophicus*. Kegan Paul. Quotations from: L. Wittgenstein, *Tractatus Logico-Philosophicus*. Mineola, NY: Dover Publications 1999.

Wittgenstein, L. (1953). *Philosophical Investigations*. Blackwell.

Woodward, R. (2011). Truth in Fiction. *Philosophy Compass, 6*(3), 158–167.

Zahavi, D., & Kriegel, U. (2016). For-Me-Ness. What Is It and What It Is Not. In D. O. Dahlstrom, A. Elpidorou, & W. Hopp (Eds.), *Philosophy of Mind and Phenomenology: Conceptual and Empirical Approaches* (pp. 36–56). Routledge.

5

The Embodied Philosopher

5.1 From Erotetic Intuition to Embodied Action: An Overview

The aim of this chapter is to provide a simplified model of the capacity to problematize. It may be attractive to those who were initially estranged by the intuition talk, for here I will not make use of the notion of intuition at all. To be clear, while traditional erotetics is concerned with the logical and linguistic aspects of questions, not with the cognitive faculty responsible for producing them, we had to nonetheless incorporate this faculty (as a black box) and give it a name. Now, if one claims that questions and their desiderata can be explained in purely logical terms, meaning that the whole area of problematization is generated by certain procedures involving propositions, then one can avoid the intuition talk and consider the whole theoretical construction to be presented here unnecessary. However, at some point we came to the conclusion that there was a level of erotetic functioning that could not be accounted for in terms of operations on propositions—a level at which there must be a direct, yet specifically erotetic, link to the questioner's surroundings. To

K. Werner, *The Embodied Philosopher*, New Directions in Philosophy and Cognitive Science, https://doi.org/10.1007/978-3-030-79964-9_5

understand this link, we need to switch to a different theoretical context, a different vocabulary.

On the other hand, this chapter enters a territory and makes use of a terminology that may estrange more traditionally minded philosophers, too. I mean, first, a paradigm in the philosophy of mind and perception—generally speaking, in the philosophical foundations of cognitive science—that has emerged over the last 30 years or so, known as embodied cognition or enactivism, among other labels. Second, I shall also refer to the more or less related area of investigation into the nature of cognition in the philosophy of biology. However, I wish to stress is that while I draw on both of these intellectual currents, I do not think that this work fits either of them exclusively. First, my intention is not to join the anti-content crusade that many enactivists who follow the radical enactivism of Hutto and Myin (2013) are engaged in (also Noë, 2004, 2009). They make a valid point, arguably, that capturing the most fundamental level of cognition in terms of content and information processing is not a good strategy, but in my view the information-processing approach can well be redesigned at some level. On the other hand, the claims I shall be advocating here could likely be rebutted by some philosophers of biology, for I share the enactivist focus on cognizer–environment (mind–world) co-determination.

For these reasons, I shall use the term "biological externalism." So in this chapter, I wish to explain what our erotetic capacity is, how it fits into a broader picture of cognition, and finally how the pursuit of philosophizing, targeted as a special exercise of our erotetic capacity, should be seen in this context.

Chapter 4 already dealt with the specificity of philosophy. This chapter provides a certain reinterpretation of what has already been said about the recognition of one's epistemic position—this time in terms of embodied action.

5.2 Mind and World: A Boundary-Conceptual Perspective

Before we plunge into a proper investigation, I have to make a remark about the perspective from which enactivism and biological externalism more broadly are viewed in this chapter.

To begin, note that every manner in which we conceptualize the mind–world relationship is *a choice* (not an individual choice but more like a choice made within whatever paradigm happens to be prevailing at a given moment, in a given culture, etc.) of specific boundary concepts in the sense laid out in Chap. 4—concepts that elicit and spell out the most fundamental "architecture" of the relationship in question. We rarely pay attention to the fact that it is, indeed, a choice to be made, and that at least one major change in this respect has occurred in the recorded history of ideas: the change attributable particularly to Descartes and Galileo. There are several ways to articulate it. Hilary Putnam speaks in this context of the "epistemic ideal of achieving a view from an 'Archimedean point'—a point from which we can survey observers as if they were not ourselves, survey them as if we were, so to speak, outside our own skins" (Putnam, 1990, p. 17). This ideal results in "the famous picture, the dualistic picture of the physical world and its primary qualities, on the one hand, and the mind and its sense-data, on the other, that philosophers have been wrangling over since the time of Galileo" (Putnam, 1987, pp. 6–7).

This dualistic conception resulted in a pivotal ontological divide between two spheres or even kinds of being: the internal realm of the mind, and the external world. From that point on, the central philosophical conundrum became the question of how to get out of the solitary confinement of the inner realm, and get a glimpse of the external reality which was conceived of as the only genuine reality, as opposed to mere projections in our heads (actually, the biggest problem was whether there *is* any external reality). And this marked a striking difference as compared to pre-modern worldviews. For, as Deborah Brown says, a

> commonly shared intuition among medieval philosophers was that the defining function of cognition was to bring the world and the mind into some kind of union, to bring the world in some fashion into the mind so as to make it intelligible (…). The soul's sharing through sensation the same forms as inhere naturally in material things enables it to know their natures. Descartes's theory of sensation is revolutionary in many respects, but perhaps most so because it puts to the test this assumption that the senses deliver to the mind the modes of being of natural objects, and finds it wanting. (Brown 2008, p. 196)

The point I wish to make is that there may be different ways of thinking about the mind–world relationship at the most general, "architectural," level. The "architecture" that we owe primarily to Descartes is based first of all on the distinction between *the internal* and *the external*, which then gives rise to a specific conceptualization of the *subjective* vs. *objective* distinction. On this basis, we tend to approach experiences as somewhat "less real" than the physical stuff out there—as epiphenomenal. In contrast, ancient and medieval approaches sought some kind of unity between the intellect and the intelligible world.

It is not the goal of this investigation to dig deeper into these historical matters. The crucial takeaway for us is this: obviously aware of all the benefits brought by the Cartesian approach, we don't have to stick with it forever. Like any theoretical posit, it does have certain limitations, and therefore we are free to search for alternatives. Biological externalism is positioned here, quite in line with the intentions of some of its pioneers such as Humberto Maturana and Francisco Varela, as such an alternative. Its crux is, in a sense, a re-introduction of the mind and its products (experiences, deemed merely subjective in the Cartesian sense) into the real world—that is, into the environment in which the mind evolved, in which minded creatures operate; and a re-introduction of the real world as a fundamentally intelligible domain.

5.3 Being-in-the-World: Rejecting the Cartesian Dichotomy

Let us begin by paying attention to one of the most *seemingly* trivial things one could think of—as used to be the practice of philosophy according to our earlier overview in Chap. 4—namely to the special locative relation of *being-in-the-world.* Being-in-the-world was made into a subject of philosophical reflection by Martin Heidegger (1982) and Maurice Merleau-Ponty (1962). Many years later it was introduced into Anglophone philosophy of cognitive science by Hubert Dreyfus (1991). Being-in-the-world is more than being placed or contained somewhere; it is more like location plus mutual dependence. Here is Merleau-Ponty's take on it (already cited in Chap. 4):

> The world is inseparable from the subject, but from a subject which is nothing but a project of the world, and the subject is inseparable from the world, but from a world which the subject itself projects. The subject is a being-in-the-world. (Merleau-Ponty, 1962, p. 430)

> The body is the vehicle of being in the world, and having a body is, for a living creature, to be intervolved in a definite environment, to identify oneself with certain projects and be continually committed to them. (ibid., p. 82)

Depending on whether we begin reflecting on the relation of being-in-the-world from the angle of the experiencing self or the angle of an external observer (such as a scientist), we come to two possible formulations of the recognition at stake: one spelled out in the first-person meditational stance championed by Descartes and calling for appropriate phenomenological treatment; and one spelled out in the third person, calling for an account in ecological and, more generally, biological terms.

That said, what seems to be unique about the embodied cognition movement is that it takes these two approaches and perspectives as equally valid, thus drawing inspiration from two sources: biology on the one hand, and phenomenology on the other (see Varela, 1995; Rowlands, 1999; Di Paolo 2005; Thompson, 2007; De Jaegher & Di Paolo, 2007;

Chemero, 2009; Stewart et al., 2010; De Jesus, 2015). It takes cognition, including more sophisticated intellectual endeavors, as a biological phenomenon, while at the same time taking experience—one might be tempted to say subjective experience—seriously, instead of deeming it a secondary epiphenomenon, reducible to its physico-chemical basis. Therefore, biological externalism can be seen as a genuinely non-reductionist (perhaps even anti-reductionist) type of naturalism.

The pivotal moment in the development of the embodied mind paradigm, and therefore also one of the pivotal points in the emergence of the broader tent of biological externalism, was 1991 when Varela, joined by his student, now a major figure in the enactivist movement, Evan Thompson and the renowned psychologist Eleonor Rosh, published the book *The Embodied Mind*. There they begin from putting into words a certain fact they refer to as fundamental circularity:

> A phenomenologically inclined cognitive scientist reflecting on the origins of cognition might reason thus: Minds awaken in a world. (…) We reflect on a world that is not made, but found, and yet it is also our structure that enables us to reflect upon this world. Thus in reflection we find ourselves in a circle: we are in a world that seems to be there before reflection begins, but that world is not separate from us. (Varela et al., 1991, p. 3)

This is nothing other than a recognition of our (human) epistemic position, as it was described in Chap. 4 (where this fragment was already cited); our position *qua scientists and philosophers* pondering about the mind. Yet this recognition cannot come about without huge ramifications when it comes to the attitude we, the readers, are invited to, perhaps even obliged to, adopt. For doing science from an alleged objectivist standpoint, as if there were no experience underneath even the most abstract models delivered by science, is not an option. Still, plunging instead into a first-person Cartesian-style meditation, perhaps in the way Husserl seemed to be doing in his *Ideas* (Husserl, 1913), is out of the question, too. That said, Varela and collaborators are open about their philosophical commitment:

> The treatment of the world as pregiven and the organism as representing or adapting to it is a dualism. The extreme opposite of dualism is a monism. We are not proposing monism; enaction is specifically designed to be a middle way between dualism and monism. (ibid., p. 202)

We are therefore somewhat pulled apart by two opposing forces: one is the tendency to focus on the "inner self," for "it is also our structure that enables us to reflect upon this world" (ibid., p. 3), and on the other hand to focus on objective facts, for we "reflect on a world that is not made, but found" (ibid., p. 3). Yet, being torn apart this way, we cannot take any of the opposite routes; thus, we cannot take any of the two sides as pregiven, take it for granted in disregard of the other.

Recall in this context the declaration made by Carnap, cited in Chap. 4, that standing on neutral ground—neutral to the *self-world* distinction—requires "bracketing" not only of the world side, since what counts as "a world" has to be logically "reconstructed," but also of the self side (egocentricity). Therefore, we can see how recognizing the fundamental circularity within biological externalism necessarily leads to some sort of "bracketing," even if it is not described exactly that way.

In effect, given the fundamental circularity adopted—noteworthily, not on speculative but on purely naturalistic grounds—Varela, Thompson, and Rosh openly claim that we have no other option but to concede that both the cognized world and the cognizer co-determine each other in a certain peculiar process called sense-making or enaction:

> cognition is not the representation of a pregiven world by a pregiven mind but is rather the enactment of the world and a mind on the basis of a history of the variety of actions that a being in the world performs. (Varela et al., 1991, p. 9)

All that said, putting stress on being-in-the-world means that we break with the long Cartesian tradition of categorizing the mind–world relationship primarily in terms of external vs. internal. This means that cognition is no longer conceived of as something happening inside our minds or brains only, which also (if we are not brains in the vat) *happens to* be placed in some kind of surroundings. Instead, cognition is *constituted*, thus made what it is, by interactions with entities in the surrounding world; it is necessarily placed in some environment.

Someone might say that the issue is purely verbal—nothing really changes when we emphasize the constitutive nature of being-in-the-world, meaning that emphasizing this particular relationship does not justify a major paradigm shift as advocated by the enactivists. For there is nothing that could a priori prevent a more traditionally minded cognitive scientist from studying the actual representational machinery operated by organisms as co-constituted by the environment they are supposed to represent (see e.g. Miłkowski, 2017). It is therefore possible for the Cartesian approach to adopt some sort of openness to the environment.

I agree that there is no need to throw the baby out with the bathwater. This is why I came up with the provisional label "biological externalism" instead of fully jumping on the enactivist platform. Yet, arguably, there is a significant point that enactivism does bear, while Cartesianism in all of its instantiations cannot. In other words, there is a pivotal novelty attributable to the embodied mind paradigm, at least as rendered by Varela et al. (1991). The very concept of an environment is what really makes a difference here. The Cartesian philosopher understands it as a pre-given realm filled with a fixed number of ready-made things that we can more or less successfully refer to, as if jumping out of our inner selves (by means of intentional acts, some would add). In other words, the environment is what it is independently of who or what perceives it, thus independently of any entities we might suppose to be cognitively open to it. That said, the rejection of this notion of environment is exactly the crux of Varela's call:

> The key point, then, is that species bring forth and specifies its own domain of problems to be solved (…); this domain does not exist "out there" in an environment that acts as a landing pad for organisms that somehow drop or parachute into the world. Instead, living beings and their environments stand in relation to each other through *mutual specification* or codetermination. Thus what we describe as environmental regularities are not external features that have been internalized (…). Environmental regularities are the result of a conjoint history, a congruence that unfolds from a long history of codetermination. (Varela et al., 1991, pp. 198–199)

Figuratively speaking, an environment is not just a place in which a cognizer is placed, like a visitor landing on it, or like a pair of shoes put into a wardrobe. Instead, an environment is a realm that is *relative to*, or *suited*, as it were, *to* the purposes, needs, and capacities of a particular inhabitant. This, however, does not mean that it is a mere projection of the Cartesian mind, made of sense-data. No—it is as real as can be. The evolutionary biologist Richard Lewontin, who by the way maintains no connection to the enactivist movement, makes a very similar point:

> There is a confusion between the correct assertion that there is a physical world outside of an organism that would continue to exist in the absence of the species, and the incorrect claim that environments exist without species. The earth will precess on its axis and produce periodic glacial and interglacial ages, volcanoes will erupt, evaporation from oceans will result in rain and snow, independent of any living beings. But glacial streams, volcanic ash deposits, and pools of water are not environments. (Lewontin, 2000, p. 48)

Therefore, Varela and collaborators insist that "our cognition is directed toward the world in a certain way: it is directed toward the world as we experience it. For example, we perceive the world to be three-dimensional, macroscopic, colored, etc.; we do not perceive it as composed of subatomic particles" (Varela et al., 1991, p. 52). The world's being a certain way for us means that it becomes a place with some bearing on our lives. Hence, as Thompson writes in one of his later coauthored papers, "even the simplest organisms regulate their interactions with the world in such a way that they transform the world into a place of salience, meaning, and value—into an environment (…) in the proper biological sense of the term. This transformation of the world into an environment happens through the organism's sense-making activity" (Thompson & Stapleteon, 2009, p. 25).

Therefore, being-in-the-world is not only about location. To put it differently, the locative relation of being-in stands here for a much more subtle and rich bond than mere containment. Yet the exact nature of that bond is still somewhat mysterious, depending on how we grasp the metaphor of the pre-given landing pad. Some more work needs to be done to unpack it more technically.

5.4 The Need for Embodied Erotetics

One might ask why it all matters when it comes to the primary goal of this book, which is a proper conceptual grasp of our erotetic capacity—the capacity to problematize in general, and the pursuit of philosophizing in particular. A short answer is this: provided that an act of problematization targets specific entities in the environment of a given cognizer, it matters a lot how the concept of environment is understood. For if we take desiderata as requests for more information about the environment, and the latter is understood as a pre-given realm of "external" and fully "mind-independent" ("not internal to the mind," in the Cartesian tradition) entities, then there is, first, a fixed amount of information to be gleaned from it (information is thought of as a kind of resource), and second, once we assume that a question has been rightly answered, it is answered for good, so that it makes no sense to come back to it. This view of our erotetic pursuits underlies the linear conception of the growth of knowledge and of progress in science, which is—sadly, according to critics—absent in philosophy. For philosophy seems to be continually dancing around the same issues since time immemorial. It would merit a separate book to investigate how this linear approach to progress proves problematic if we look at the actual history of science itself, so I shall not discuss it here. Instead, just a few quick remarks in relation to the pursuit of problematization.

First, if the problematized environment is not pre-given, each setup of questions posed at a given moment is relative to the relevant state of the continuing process of sense-making, as Varela et al. (1991) put it, and it is actually part of that process. In other words, problems are attributable to a given moment in the evolution of environments, and once environments change in some of their aspects, some seemingly identical questions must be posed anew, for they are actually not the same questions at all. Now, this does not mean that there is no progress or accumulation of knowledge—these two correlated ideas can be re-introduced in this new context; however, their sense will clearly be different. Nor does all of this mean that there is no right answer to a given question, clearly distinguished from wrong answers. I am not trying to make that point. Rather,

I would like to explore how the pursuit of problematization actually works as part of sense-making and how it emerged on the basis of the actual processes that living creatures are constituted by and involved in.

Embodied erotetics, if we can dub it thus, seems a rather underdeveloped field, to say the least, not to mention embodied meta-philosophy, understood here as a specific application of erotetics. I aim to fill this gap, at least a tiny part of it.

5.5 Cognition as a Biological Phenomenon: The Life–Mind Continuity Thesis

In biological externalism, the cognizer is approached as a living creature, an organism.[1] A non-trivial aspect of this somewhat obvious claim soon shows up, however, for if cognition does not boil down to internal computation, but is instead part of living, it must by definition contribute to the process of maintaining the integrity and survival of the living creature. While pondering how cognition contributes to survival and evolutionary success has become quite fashionable in the field of evolutionary anthropology (see Dunbar, 2014; Tomasello, 2019), among other places, the former puzzle goes much deeper than that. Here, people such as Maturana and Varela (1980) or more recently Alvaro Moreno and collaborators (1997), Marc van Duijn and collaborators (2006), Peter Godfrey-Smith (2016), Fred Keijzer and Argyris Arnellos (2017), among others, aim to fathom the very core of life, nail down what it means to be a living creature (also see Kirchoff 2016; Cummins & De Jesus, 2016). There, they expect to also find some hints as to what it means to be a cognizer. This suggests a close link not only between survival and cognition, which is widely acknowledged, but also between life itself and cognition. This view is known in the embodied cognition literature as the *life–mind continuity thesis*, and it was reintroduced by Maturana and Varela with their papers on autopoiesis. I say "reintroduced" since it was adopted already by Jakob von Uexküll (1926) as well as the pioneers of comparative psychology such as C. Lloyd Morgan (1894) and Margaret Washburn (1908).

Autopoiesis literally means *self-production*. An organism is approached as a complex system whose unique capacity, absent in other natural complexes as well as in artificial, man-made systems, is the one thing that enables it to maintain its own boundedness, its own integrity, by re-producing its own building blocks (see Maturana & Varela, 1980; Weber & Varela, 2002).

Autopoiesis is linked to another notion widely cited in the literature: that of autonomy. Since autopoiesis is supposed to capture the essence of the simplest living form—the unicellular organism—autonomy is regarded as a more general notion, applicable to both relatively simple and relatively complex organisms, including humans.

Summarizing the literature on the subject, we can say that:

(21) For a given complex entity to be an autonomous system, or to maintain its precarious autonomy, the following conditions must be satisfied:

(a) The entity is realized in a fixed domain or medium—it has certain components.
(b) Its components recursively depend on one another in their (re) production as components of the system.
(c) It is bounded and maintains its boundaries.
(d) It determines a domain of possible interactions with entities that surround the system.

Now, the somewhat obvious claim that cognizers are living creatures leads to the highly non-trivial posit that cognition is targeted as a contribution to, and at the same a development of the most basic and fundamental processes maintaining the living creature: its autonomy. This means that even if living itself does not mean cognizing, as Maturana and Varela's (1980) original approach suggests, the connection nevertheless goes deep and one virtually cannot properly understand cognition without seeing it in light of autonomy.

What applies to cognition in general applies to the pursuit of problematization in particular, including the pursuit of philosophical problematization. Targeting the philosopher as a living creature maintaining its autonomy may seem outrageous to some, but let us keep in mind that this is only a simplified heuristic that is supposed to open some new prospects for meta-philosophical reflection, not a complete reductionist picture of philosophizing as nothing more than taking care of our physiology. It is not my intention to make the latter point.

5.6 The Niche

The conditions of autonomy determine which parts or aspects of the world are relevant, thus which are required for the system to maintain its precarious autonomy. Here we come to a first approximation of another crucial notion—*the niche*.

Beginning with the works of Lewontin (1983), carried on by Kevin Laland, among others (see Laland et al., 1999, 2000; Odling-Smee et al., 2003; Laland & Sterelny, 2006; Sterelny, 2007, 2012; also Ryan et al., 2016 and Wallach, 2016 to discuss wider applications of the concept), niche construction theory has become one of the major paradigms in contemporary evolutionary and theoretical biology, effectively working as a constraint imposed on the selectionist approach to evolution (see, also, e.g., Gould & Lewontin, 1979; Godfrey-Smith, 1996). Quite rarely used by enactivists, the concept of niche in turn gives us a chance to unpack, less metaphorically and instead more technically, some of the intuitions regarding being-in-the-world and the enactivist idea of the environment that is not pre-given but enacted. Of course, this new application of the concept brings major changes to its meaning, especially when compared to the way in which it is usually understood by biologists. That said, the new elements shall be introduced stepwise, so that we can trace a certain evolution of the concept instead of making an abrupt transplantation of it from one theoretical context to another.

To begin, likely the most intuitive characterization of a niche, pioneered by Charles Elton (1927), portrays it as a specific place, a *locus* occupied by some living creature, for example, a woodpecker's hollow. One may ask, however, what makes a particular place special to the organism. The instant answer would likely point to two elements: first, the locus in question is usually transformed by the organism in some way; and second, it provides the organism with the resources it needs to thrive.

Beginning with the first element, it hints at a crucial issue: niches are constructed, so speaking of niches is in fact speaking of the processes of *niche construction*, setting the specifics of this constructive pursuit aside for now. From the perspective of an evolutionary biologist, this claim is pivotal, since it introduces niche construction as an additional constraint on selective pressures, an additional evolutionary factor (Lewontin, 1983).

When it comes to the second element, providing and securing resources, it literally means that the organism is capable of, for example, storing food in its nest or burrow, yet the idea can easily be generalized. It is not only about food and water but about any *conditions* that the surrounding world may or may not satisfy, be it the availability of food, but also temperature, air pressure, humidity, and so on.

Given this more general approach, we can modify our understanding of the niche, following Lewontin, who said that "ecological niche is a technical term universally used to denote the complex of relationships between a particular species and the outside world" (Lewontin, 2000, p. 44; see also Lewontin, 1983; Laland et al., 1999), thereby moving away from the locative model toward a more abstract, relational, one.

As for the more abstract approach, yet in a different rendering (which has already become classic in the field), we have to refer to G. Evelyn Hutchinson (1957), who pointed out that if we take into account any two "ecological factors" x_1, x_2 relevant to a given organism *o*, for example, temperature and the availability of certain nutrients, then measure them along the Cartesian coordinates and determine some limit values for these factors, we define a certain "space" within the system of coordinates. This "space" is an abstract characterization of the niche required by *o*. This gives us an idea of what a more general view of the niche could be. For, if we consider all the ecological factors $x_1, \ldots, x_n$ relevant to *o*, adding many new dimensions to the system of coordinates, we obtain an *n*-dimensional system such that every point in it represents a fixed state in *o*'s surroundings, and *o*'s niche can be described as a set of points representing all and only those states that enable *o* to thrive.

Hutchinson's characterization enables us to free ourselves from the limits of our imagination and conceptually capture also the less straightforward, but still genuinely real-life examples of niche-like structures. In this vein, recent literature speaks of, for example, the cultural niches of complex animals, the long-lasting, transgenerational character of which proves to be another major factor in evolution (see e.g. Sterelny, 2012).

To resolve the tension between the abstract and the locative approaches, Barry Smith and Achille Varzi (1999) propose the following two distinctions: first, a niche *token* is a fragment of the world, thus a particular locus, while a niche *type* is not individuated by its place but by conditions to be satisfied. The former stands for the niche in Elton's sense, and the latter unpacks Hutchinson's proposal. The two posits perfectly complement each other. For at the end of the day, the abstract conditions posited by Hutchinson have to be realized somewhere, thus a niche type always needs to be instantiated by a niche token. Moreover, they distinguish a niche *medium* and a niche *base*. The latter refers to the material (actually the type of material) out of which a niche can be built, for example, moistened soil in the case of a worm's burrow, and twigs in the case of a bird's nest. The niche medium, on the other hand, is what *actually* serves to make a given niche, so not all potentially useful things, not even all the twigs in the world, but those collected by a particular bird in a particular forest. More generally speaking, the talk of niche medium brings, in a sense, a more "subjective" perspective, that of a niche constructor, into the picture, while the talk of niche base is more in line with the external observer's perspective. The medium is more like the base *as seen* from the perspective of the animal. To put it in Heideggerian terms (see Heidegger, 1962), while speaking of the base refers to what is *at*-hand, meaning that it is described in disregard of what are a particular niche constructor's needs, speaking of the medium is more like referring to the material that is *ready-to*-hand. The latter refers to things that are relevant to the niche constructor's actual interests and capacities.

We are now finally on the right track to operationalize the Heideggerian relation of being-in-the-world: in biological terms, it means for a living creature, including humans, to construct and to be placed in its own niche. We need a further step though, for we still have not said anything about cognition, not to mention the pursuit of problematization which is our major concern.

5.7 The Cognitive Niche

I will now make use of the abstract account pioneered by Hutchinson and developed by contemporary niche construction theorists. For instance, Kendal et al. (2011) stress that "the theory incorporates cultural evolutionary processes" and as such it "provides a framework for consilience between the natural and the social sciences" (Kendal et al., 2011, p. 785; see also Wallach, 2016 for a more critical approach). However, a missing step between the two sciences, I would insist, is a niche-theoretical account of cognition itself, which is needed before we move up to the most sophisticated products of cognition such as culture.

To begin, there is no a priori limit as to what conditions can in principle define a niche and what conditions cannot. It seems more realistic to avoid drawing artificial divisions within the things that decide on a given organism's living and survival; things that may have to do with many different aspects of the organism's functioning and needs at the same time. Think of the following example: suppose that a worm enters the nest of a bird, thus an edible fellow animal. The worm responds to the bird's need for food, yet in order to actually be eaten it must be perceived first and recognized by the bird as a potential lunch. Then, the bird must exercise a series of movements and sensorimotor coordination functions in order to capture the worm, and this also imposes certain conditions on the bird's surroundings. If, say, everything is going on in an animal behavior lab and the worm is in fact behind a perfectly transparent glass pane, the conditions do not enable the bird to exercise its sensorimotor capacity of pecking properly. Note that while the worm is directly linked to the bird's metabolic needs, its *presence* must be made available, as it were, by the bird's cognitive tools, which in turn sets up a different set of conditions in the animal's surroundings. Now, take an example exploited by Smith and Varzi (1999), that of a bear's cave. A bat may reside in the very same cave, and moreover, the bat may be seen by the bear, yet—unlike in the worm and bird example—this cognitive link does not mediate between the metabolic need for food and its satisfaction, since the bat is completely irrelevant to the bear's needs of any kind. Yet the bear can still *see* its roommate. Both of these cases—the one in which cognition serves as a means to execute certain actions and satisfy a certain metabolic need, and especially the one in which cognition is just on its own, as it were, serving no greater purpose—show that we can think of a specific realm around the living creature that is defined exclusively by that creature's *cognitive* capacities (and also its cognitive needs, if this notion makes sense), as distinguished, for example, from its purely metabolic needs. I propose to refer to this specific realm by a term first proposed in a slightly different context by anthropologists John Tooby and Irven DeVore (1987), namely "cognitive niche."

Tooby and DeVore (1987) set out to identify the specificity of human cognition. According to them, humans have a unique "ability to create and maintain cause–effect models of the world as guides for prejudging which courses of action will lead to which results" (Tooby & DeVore, 1987, p. 2010). The point is not that humans are highly flexible in their behavioral capacities (which is true, of course), but rather that they manifest more sophisticated capacities to select their actions in accordance with specific needs or goals. Since "there are an infinitely large number of possible sequences of behavior (…) 'behavioral flexibility' is an insufficient characterization of our innovative adaptive pattern. Our cognitive system is knowledge or information driven, and its models filter potential responses so that newly generated behavioral sequences are appropriate to bring about the desired end" (ibid.).

The notion of cognitive niche gained some traction due to Steven Pinker (2003, 2010) and Andy Clark (2005, 2006) in particular. Somewhat similar is the concept of epistemic niche developed by Stotz (2010). In recent years there have been several works by Bertolotti and Magnani (2017), and myself (Werner, 2018). Noteworthily, the concept at stake brings to mind the classic concept of Umwelt introduced about a hundred years ago by Jacob von Uexküll (1926), which can be described as a kind of information cocoon—if I can borrow that term from contemporary media studies—determined by cognitive tools, life needs, and capabilities.

Another well-established theoretical posit that can be used as a template for our intuitive fleshing out of the cognitive niche is James Gibson's (1979) concept of affordance in the domain of cognitive psychology.

Gibson is quite enigmatic and does not provide a technical characterization of the posit, but here is the one that is probably the most cited:

> The *affordances* of the environment are what it *offers* the animal, what it *provides* or *furnishes*, either for good or ill. (Gibson, 1979, p. 127)

We can say that affordances are either properties of the surrounding world or relations between the cognizer and its environment (see Chemero, 2003; Brancazio, 2020) that are somehow relevant to the cognizer, usually (if we consider their positive impact) giving the living creature in question an opportunity or a means of doing something. This clearly makes the notion relevant to the concept of niche as presented above. Interestingly, the link is quite clear to Gibson himself, who defines the niche as a set of affordances offered to the niche inhabitant.

The concept of cognitive niche proves crucial later in this chapter. First, when it comes to the ideas put forward in this book, note that one's epistemic position can be unpacked, thus in a sense "operationalized," in terms of cognitive niches, while its recognition can be portrayed as becoming aware of one's cognitive niche as such, as *one of many possible realms* of this species-specific kind of niche. When I recognize my epistemic position, I begin to see my cognitive niche *as a cognitive niche.*

Second, note that a living creature's (be it a human or any other thinkable cognizer) capacity to problematize can also be characterized in terms of cognitive niches and their components. It will be the creature's capacity to take a certain relatively stable part or segment of its cognitive niche, thus a relatively robust product of niche construction, and re-introduce it into the active process of niche construction or, as enactivists would say, into sense-making.

Both of these reinterpretations will be developed below.

5.8 Cognitive Niche: Troubles with the Base

We need to clarify the concept of cognitive niche. For there is a variety of approaches, and while this abundance is initially a positive thing, some decisions have to be made to order the field. To be more precise, we must define the base (in Smith & Varzi's, 1999 sense) of the cognitive niche, and then its medium. Of course, one might say that at the end of the day all niches are based on the same physical stuff which all of reality is composed of, and it's in some sense true, yet it is trivially true in the sense that it brings nothing of value to our current investigation, especially given what Lewontin (2000) said (cited in 5.3).

The original posit by Tooby and DeVore (1987) emphasizes internal models of cause–effect relations, the capacity of exploiting them, and—last but not least—the uniqueness of *Homo sapiens* in this respect. There are several problems with this view. First, the uniqueness of human beings, even if to some extent understandable given how sophisticated we have become in our cognitive functioning, gives little chance to learn more about how these extraordinary cognitive capacities evolve, how they emerged from more "primitive" functions.

It is not the uniqueness claim that causes real trouble here, however, but a tacit presumption characteristic of the Cartesian thinking that the modeled environment, clearly external, thus walled off from the internal domain of representations, is what it is independently of its inhabitants—that at its core it is pre-given and thus not constructed but rather inhabited by organisms (Bertolotti & Magnani, 2017 discuss various interpretations). Meanwhile, driven by the recognition of our epistemic position—by the recognition of fundamental circularity, following Varela et al. (1991)—we cannot accept the concept of a pre-given environment. And we cannot accept the gulf between the internal and the external.

Clark (2005) comes up with a slightly different account. First, he includes non-human animals among cognitive niche builders. Second, the niche is not merely inhabited—it emerges from re-constructions of the surrounding world made by the cognizer, usually to ease cognitive functioning. We can read:

> "cognitive niche construction" [is] hereby defined as the process by which animals build physical structures that transform problem spaces in ways that aid (or sometimes impede) thinking and reasoning about some target domain or domains. These physical structures combine with appropriate culturally transmitted practices to enhance problem-solving, and (in the

> most dramatic cases) to make possible whole new forms of thought and reason. (Clark, 2005, pp. 256–257)

The trouble is that although the cognitive niche comes about as a result of specific actions, the *problem* space which the cognizer operates in seems to be, again, already there, pre-given, as if independent of whether there is any cognizer able to problematize segments of its surroundings.

Another option is to unpack the cognitive niche in terms of information. It would not be in opposition to what Tooby and DeVore (1987) originally intended, and it is openly suggested by Pinker (2003). Actually, it is a widely shared view that cognition is information processing and humans are especially skilled in it. Therefore, cognitive niche theories "born in the 'cognitive/philosophical sector' of niche theories have stressed how human beings' relationship with their environment is essentially information-based" (Arfini et al., 2019, p. 379). From this angle, "gathering and exchanging information is essential to the niche that grounds the entire human ecology" (ibid., p. 380).

Noteworthily, niche construction theory more broadly—thus not only those who make explicit use of the term "cognitive niche"—also spells out the specificity of human niches in terms of information. For example, Odling-Smee (2010) discusses the mechanisms of niche inheritance and cultural transmission and holds that niche inheritance boils down, among other things, to the transmission of *information resources*.

A proponent of enactivism would protest at this point, since speaking of information may bring back the whole representational paradigm. However, that may also not be the case. For example, we read:

> Semantic information in an individual's inherited niche might take the form of a behavior, demonstrated by peers or elders, acquired through social learning, such as subsistence practices or social norms. (Kendal et al., 2011, p. 787)

The problem with information talk is not necessarily that it smuggles in representationalist baggage. It is analogous to the problem with pre-given environments and problem domains. Namely, the term "information" seems to refer to some mysterious stuff just waiting to be picked up.

This is what speaking of information *resources* suggests. To get a sense of this point, compare information with other kinds of resources. Arguably, coal is a resource—an energy source to be precise—only for the creatures who can make use of it. Yet coal is coal, however trivial it may sound, whether or not anyone extracts and uses it. Of course, the category "coal" comes from human thought and it already has a ready-to-hand aspect, as Heidegger, 1962 could say, but insofar as it refers to carbonized plant matter, everything there is to be said about the material named "coal" boils down to its chemical properties. It seems that the case of information resources is different. This is clear when we consider notions like semantic information or representational information, as discussed by Dretske (1995). For in order to serve as a vehicle of information, a particular entity must be appropriately processed by the receiver. This also applies indirectly to quantitative measures of information pioneered by Shannon and Weaver (1949). For if the amount of information carried by an event x equals the negative logarithm of the probability of x, the event x must already be specified and, at least given the fundamental circularity principle invoked earlier and the notion of organism-environment co-determination adopted by biological externalism, there is no way to individuate x in disregard of a fixed living creature (or a variety of them) enacting x as part of their environment.

All this, on the other hand, does not mean that the information talk is useless and misplaced. On the contrary, I shall go back to it and articulate my major posits at least in part in terms of information. Yet, I hold that a proper description of the emergence of cognitive niches cannot be articulated in terms of information.

What is left, I propose, is a genuinely enactivist proposal. I shall rely on Ezequiel Di Paolo et al.'s (2017) comprehensive model of cognition, centered around the notion of sensorimotor coordination, rooted in Jean Piaget's (1969) concept of equilibriation and resembling (although he is not cited in the book) Heinz von Foerster's (1960) notion of *eigenvalues*. I propose to try out the idea that the creation of sensorimotor invariants and schemes sets off processes of cognitive niche construction. In other words, cognitive niches are at their bottom level composed of relatively stable invariants of sensorimotor coordination produced by the autonomous system's interactions with its surroundings. In this sense, these sensorimotor invariants are good candidates for the sought base of the cognitive niche.

5.9 The Cognitive Niche: Introducing the Sensorimotor Base

To begin, Di Paolo and collaborators (2017) aim to explain how a relatively stable realm manifesting itself in perception comes about, without the representationalist faith that the world outside our heads is basically what it appears to be, just waiting to be pictured. Perhaps this faith is not in vain, but insofar as we cannot glimpse this alleged external world from a God's Eye perspective, we need to capture the emergence of the manifest perceptual regularities in a different way. The mentioned regularities or invariants are dubbed *sensorimotor contingencies* by Di Paolo et al. (2017)—SMCs, for the sake of brevity. There are several kinds of SMCs and since my purpose here is not to examine the proposal in question but to actually make use of it, I will briefly list what differentiates these various kinds and then move forward.

The simplest case of SMCs is represented by the simple relation of the placement of a cognizer, a creature equipped with at least some primitive sensors, within a particular causal network. A specific position of the cognizer causes, for example, the retina to be stimulated in a particular way; the cognizer's movement causes regular changes on the retina; and that's it. Nothing depends here on how the stimulation is influenced by the cognizer's neural activity as a whole, how it is connected with past stimulations, nothing depends of the cognizer's skills, and so on. We read:

> Movement of the body leads, for example, to lawful instantaneous changes in the optic flow on the retina (expanding when moving forward, contracting when moving backward, etc.). Equally, rotations of the head lead to a lawful change in the temporal asynchrony between sounds received in the left and right ear. These lawful relations are independent of how such movements originate or combine with other movements. (Di Paolo et al., 2017, p. 53)

As "primitive" as it is, this level of perceptual regularities, giving rise to what the authors call the sensorimotor environment, is pivotal for all further kinds of SMCs built onto it by adding various factors.

Further on, the first addition is the whole internal neural activity of the cognizer. This means that perception at this stage depends not only on what happens at the surface of the sense organs, for example the retina, at a given moment, but on the entirety of the neural processing which is able to connect the current stimulation with previous ones, thereby enabling the cognizer to actually recognize regularities or patterns. The perceptual realm which thus emerges is the *sensorimotor habitat*.

The regularities in sensorimotor habitats may be quite complex, yet they abstract from what the cognizer actually does—from its goals and tasks. So while both the sensorimotor environment and the sensorimotor habitat placed the cognizer solely within networks of causal relations, external and then internal, the next step is to introduce a functional network: one that is not fully determined by lawful causal bonds, but also by the cognizer's performance of specific tasks. Here Di Paolo and collaborators speak of sensorimotor coordination. If we take the example of a missile equipped with a guiding system explored by O'Regan and Noë (2001), we can distinguish between what the missile's surface sensors register (the sensorimotor environment); how the guiding system reacts to changing inputs, how it combines them, and how it changes the "behavior" of the missile (the sensorimotor habitat); then—finally (the level of sensorimotor coordination)—we can take into account the actual target and how the above processing serves to achieve a particular task, for example, the missile hitting a particular location.

Finally, the last element of the hierarchy emerges when we take into account not only the causal + functional network but also the normative one in which certain actions may be skilled to a greater or lesser extent, thus contributing to the task in varying degrees. This realm of regularities is referred to as *sensorimotor schemes*. In short:

> A normative framework involves reference to given criteria that distinguish or value some possible outcomes as preferable to others: a dexterous movement versus a clumsy disaster, achieved know-how versus lack of expertise, and so on. (Di Paolo et al., 2017, p. 57)

Given the latter, our investigation has just come full circle; we have come back to the issue of autonomy. For if someone asks what constitutes the normative network completing the structure of perceptual regularities, and what determines the nature of an organism's tasks, the answer is, arguably, the specific conditions of autonomy.

Now, Di Paolo and collaborators propose a semi-formal characterization of perceptual invariants in terms of functions. Suppose that the variable e ranges over possible states of the environment; s refers to states of the sensor activity (e.g., a state of the retina) and a to states of further neural activity; finally that m represents possible states of motor activity. All of these variables are, of course, closely correlated—motor activity obviously depends on the position of the body and muscle contraction; the neural activity of the central nervous system obviously depends on what is going on in its peripheral sight, and vice versa. Some of these correlations—those that are significant to the emergence of perceptual invariants—are represented in the form of the following functions. First we have two binary functions, S and A:

The rate of change of the sensory activity, thus s, is a function of the states of the environment and the states within the neural circuits:

(22) $(ds/dt) = S(e, a)$.

The rate of change of $\boldsymbol{a}$ depends on the states of $\boldsymbol{a}$ itself and on $\boldsymbol{s}$:

(23) $(da/dt) = A(a, s)$.

Finally the rate of change of the motor variable depends on $\boldsymbol{a}$, and can thus be represented as:

(24) $(dm/dt) = M(a)$.

The above functions represent the internal functioning of the cognizer, including what is going on at its surface (in the sensory organs), as well as the direct outcome of these internal dynamics in the form of the cognizer's motor activity. Yet for the purposes of this model all of these factors are in a sense abstracted from the actual cognizer as an embodied whole.

Therefore, another variable, b, ranges over possible states of the body (its position, configuration, muscle activity, etc.). This is a pivotal part of the entire story, for the following two functions involving b are supposed to formally unpack the open-ended loop between the cognizer and the environment. First, the rate of change of the states of the body is a function (B) of these very states, the cognizer's motor activity and the environment put together:

(25) $(db/dt) = B(b, m, e)$.

Yet, the rate of change of states of the environment is, crucially, a function (E) not only of the said states, but also of states of the cognizer's body:

(26) $(de/dt) = E(e, b)$.

This latter point is most pivotal, since had it not been there, one could accuse Di Paolo and collaborators of taking the environment as a pre-given realm which is just there, independently of the cognizer.

Now, having set all of these variables, reciprocally determined by one another, we can represent invariants of at least two types. First, according to Di Paolo and collaborators, the sensorimotor environment can be represented by the partial derivative $\partial s/\partial m$ which stands for change in the values of s resulting from change in m, while m is taken as an independent variable and all other variables are taken as constant. Lawful, relatively robust correlations between changes on the sensory side and the relevant changes on the motor side unpack the idea of a perceptual invariant at the most basic level.

Sensorimotor habitat is in turn represented by a more complex functional structure, namely the total derivative ds/dm, where m is no longer an independent variable but a variable dependent on other listed variables, and therefore all the above functional correlations between s, m, a, e, and b apply.

All that being said, I propose that sensorimotor habitats, as understood by Di Paolo et al. (2017), can be regarded as the sought bases of cognitive niches. Once a certain sensorimotor habitat emerges, relative to a specific autonomous system or a species thereof, provided that the system's conditions of autonomy are fixed, the process of constructing a cognitive niche can begin. But how does it actually happen? Once we have the base, which was the pivotal thing to find, we can re-introduce the talk of models used by Tooby and De Vore, as well as the information talk, but this time without the Cartesian/representationalist presumptions.

5.10 Cognitive Niche Construction: A Need for an Ontology

Someone might ask about the two further levels distinguished by Di Paolo et al. (2017), namely sensorimotor coordination and sensorimotor schemes; how they fit in, and why the niche talk is to be added in the first place. My answer consists of several steps. First, note that the last paragraph refers to conditions of autonomy—they determine the needs and goals of a living creature as if answering the question: "So now, once I have certain perceptual invariants, thus I am surrounded not with a flux of stimuli or clouds of physical particles (as a physicalist might say) but with remotely stable perceptual invariants relative to my body, what am I supposed to do with all this?" In this context, here is my major point: the conditions of autonomy define the kind of place that the organism must build for itself *based on* the realm of invariants. Thus autonomy defines the Hutchinsonian niche of the organism in question.

Therefore, the "functional network," as Di Paolo and collaborators call it, that is, a realm that is not fully determined by lawful causal bonds, but also by the organism's performance of specific tasks, can emerge, in my view, once the organism starts to make use of its surroundings, which means—*when it starts to construct its cognitive niche*. Put differently, sensorimotor coordination is here taken as part of cognitive niche construction. However, it was the emergence of sensorimotor habitats that was truly pivotal in the sense that it laid the groundwork for the niche-constructive activity, providing the material—the base—that the cognitive niche is built from (some invariants must be there to begin with).

Regarding the normative framework as seen by Di Paolo, that is, sensorimotor *schemes*, I claim that once the niche construction is underway, it *begins* to make sense to distinguish skilled and less skilled executions of the living creature's capacities. In other words, in my view the normative aspect of the living creature's actions also emerges in the context of that creature's pursuit of niche construction.

However, the interlocutor who brought up the question about sensorimotor coordination and schemes may continue this way: why are we supposed to introduce new notions—cognitive niche and niche construction—to the conceptual framework proposed by Di Paolo and collaborators if their framework captures all the significant moments of the constructive process we are talking about?

In order to answer the latter concern we have to hark back to the way we experience the world around us, and to the experience of ourselves in it, which is a maneuver we are allowed to perform based on an assumption rolled out earlier. Namely, part of the phenomenological legacy within enactivism and, I would insist, biological externalism more broadly, is that experience is a fully legitimate source of knowledge about who we are as living creatures and cognizers. For at the end of the day this is the most *concrete* thing we have, composed not of theoretical postulates but, we are tempted to say, real down-to-earth stuff that we see, feel, take pleasure in or suffer pain from, and actually deal with when managing our lives. With the latter in mind, recall what I wrote a moment ago: that the conditions of autonomy define *a* place that the organism must build for itself based on the realm of invariants. If this statement did not spark controversy, it is partly because we already introduced the fundamental locative relation of being-in-the-world, and the purpose of the talk of invariants, schemes, and so on is ultimately to unpack the meaning of this relationship. Yet, unlike the world that we are *in*, the talk of invariants, schemes, and so on does not involve any locative notions. The term "habitat" used by Di Paolo and collaborators is somewhat misleading in this context, for while it has a locative aspect, the latter is left completely unexplained by the authors.

Therefore, in my view what still lacks theoretical treatment is the emergence of actual *objects*—trees, buildings, cars, mountains, and clouds; actual *processes* such as water flowing in a mountain stream; actual *events* such as someone's abrupt entrance into a room, and actual *places* where all this comes about. Note the difference: what we have is a story about sensorimotor bonds, for example, we have a bond between the noise made by a person entering the room on the one hand, the neural circuits connecting this stimulation to previous ones, and the connection between the stimulation and our skillful response like covering our ears to protect them. But wait a minute, I just said "the noise made by a person," while in our theoretical model we are still lacking the noise *made by the person*; we are still lacking that person's *entering the room*. What we have are perceptual invariants at various levels, with varying degrees of complexity and cognizer-surrounding engagement, but the familiar beings we are surrounded by are not themselves invariants even if they can be theoretically decomposed into invariants.

The question is, therefore, what the talk of objects, events, or processes (this list is somewhat ad hoc, incomplete, and can perhaps be expanded) elicits, and what the talk of invariants lacks. Here are some hints: first, the experienced beings have parts; they are bounded, thus have boundaries that they are usually limited by, with the exception of such fuzzy objects as a swarm of bees; they also occupy a certain place; each one has a specific address; technically speaking—each has its exact location.

My point here is that we need to be able to account for how the emergence of invariants gives rise to *the world as we know it*: the world of bulky entities with parts, boundaries, and addresses. Luckily, there are formal-ontological disciplines devoted to these issues. First, mereology is the formal ontology of parts and parthood developed by Stanisław Leśniewski (1916) and then by Henry S. Leonard and Nelson Goodman (1940; see also Simons, 1987; Varzi, 2007). The concept of boundary is one of the conceptual foundations of a certain cluster of mereology and topology dubbed mereotopology, founded by Barry Smith (1996) and Roberto Casati and Achille Varzi (1999; see also Smith & Varzi, 2000), which investigates not only what it means for an entity to be composed of parts, but also what it means for it to be bounded, to be a self-connected whole, and what it means for parts and wholes to be continuous or discontinuous.

Hence, although there is no flaw in the model put forward by Di Paolo and collaborators, there is a certain gap in it, at least from the phenomenological and, subsequently, ontological perspective as just described. I believe that in order to provide a viable story of how certain goals and skills emerge in the domain of invariants built upon sensorimotor habitats, we must have some grasp of *how invariants actually produce the manifested world we know*. For goals have to do with objects and events, not with invariants themselves. Objects become tools—ready-to-hand things—not invariants themselves. Think of the missile example again—however outrageous it may sound to a peace-loving heart, the goal of the person who launched the missile was not a certain sequence of percepts, but the destruction of a certain enemy target. The same applies to skills—one is not skilled in coordinating stimuli; rather, we say that one is a skilled dancer, a skilled boxer, and so on.

Now, my goal here is not to introduce the whole conceptual framework of mereotopology; we have no place for that. Instead, I will first speculate about what the construction of a cognitive niche actually entails, and then—for the sake of creating a simplified model of this constructive process—I will sketch out a kind of *ersatz* ontology, thus not a formal ontology proper, but an ontological ingredient to complement Di Paolo's functional approach. This will enable me to finally grapple with the question of how the pursuit of problematization emerges as part of cognitive niche construction and what its basic mechanism is.

5.11 Cognitive Niche Construction: In Search of a Medium

Di Paolo and collaborators (2017) provided us with a potentially useful account of the cognitive niche base. Yet it turned out that there is more work to be done if we are to account for how the actual world *qua* experienced is brought forth or enacted on top of this base via cognitive activity. For while the invariants populating the sensorimotor habitat are the base, that is, the raw material that enaction works with, so to speak, the entities that it produces are not themselves invariants—they are, as our own experiences teaches us, the familiar objects, processes, events of everyday experience. In other words, objects, processes, and events, not invariants themselves, compose the actual *medium* of our cognitive niches. There is therefore a gap that we need to fill in our theoretical account of cognitive niche construction, regarding how the base described by Di Paolo and collaborators is transformed into the actual medium of the cognitive niche.

Since there is no room here to provide a proper ontology of niche construction, we must work with a simplified story. Cognitive niche construction should be seen as generating certain clusters or fusions of invariants, relative to the cognitive capacities and needs of a given autonomous system. We can say that autonomy decides what clusters of invariants have to be glued together if the organism is to thrive. That said, putting ontology aside, we can rediscover potential in Tooby and DeVore's (1987) concept of modeling cause–effect chains as well as Clark's (2005, 2006) account of cognitive niches in terms of transformations. Beginning with the former, it must be noted upfront that while Tooby and De Vore worked within the representationalist paradigm prevailing in their time, this affiliation is not necessary. Motivated by Rodney Brooks' (1991) famous dictum that *the world is its own best model*, we can try out a more externalist/embodied/enactive approach to modeling (see Zebrowski, 2010 on Brooks and Merleau-Ponty). While Tooby and De Vore speak of selecting and exploiting cause–effect chains in the outside world, we need to speak of exploiting and altering sensorimotor invariants, which does not seem to contradict the original intention.

Keeping in mind that we have sensorimotor invariants as a base and that we are in principle at a very basic level of cognition, it might still be useful to first give a real-life illustration to get a sense of what is being put forward here. I already provided the following familiar example in Werner (2019), namely that of a bookshelf. Since, obviously, things fall due to gravity, a way to keep books upright is to put them on a shelf. Now, one way of modeling gravity is what Newton did by providing a formal model, yet a more "primitive" way of modeling gravity would be to build something that prevents the relevant causal chains from occurring. So a construction preventing books from falling down is also a "model" of gravity. This means that the cognizer's capacity to transform entities in its surroundings can also serve as a way of "modelling" the causal relations between them. Now, crucially, the story of such "models" can be spelled out in terms of observed regularities, thus certain lawful correlations between invariants within our sensorimotor habitat.

This example, however, is somewhat over the top, for it involves full-blown cognitive acts and complex manipulation. If we are to get a sense of how the medium of a cognitive niche comes about on the basis of sensorimotor invariants, we need to look for more "basic" cases. Still, the example provided above does some good, too, for it takes our thinking in the right direction—toward an idea of modeling that does not take place inside one's head, but happens as a component, or perhaps—especially in these more "basic" cases—as a byproduct of managing the precarious autonomy of the living creature against perturbations in its sensorimotor habitat. Think of the already invoked bird's nest or beaver's dam. These animals, too, select certain invariants in order to make use of them, to exploit them by actively transforming them. The word "transformation" is perhaps not the most apt, since the invariants themselves are not altered, rather the living creatures literally make an effort to cluster selected invariants as if juxtaposing them, sometimes directing, metaphorically speaking, one against another (as in the bookshelf preventing the books from falling) in order to enact yet another invariant (that the books don't fall, which becomes an invariant, too). However, birds and beavers are also full-blown cognizers, even if not as sophisticated as humans. But think of a worm digging a burrow in the soil: it is precisely due to the exploitation of simple invariants that the soil begins to *mean* anything to the worm; it becomes part of the worm's species-specific Umwelt, as von Uexküll (1926) would have said. Moreover, as Godfrey-Smith (2016) appealingly pointed out when speaking of re-afference, the exploitation of invariants (Godfrey-Smith used a different conceptual framework) when digging the cavity is also co-responsible for a primitive form of self-awareness of that worm. For it must be able to distinguish the invariants correlated with its own moves, and those independent of them ("externally" caused, thus potentially dangerous). So, "modeling" sensorimotor invariants by making actual use of them may also responsible for the most primitive feeling of self, which is in full agreement with Varela's et al. (1991) co-determination thesis.

The mentioned worm is perhaps also already a cognizer, thus what it transforms is already a cognitive niche medium, but based on the downward ladder of cognitive sophistication we can at least think, thus theoretically postulate, a level not exactly of cognitive but pre-cognitive transformation determined by a living creature's maintenance of its precarious autonomy, whose byproduct, so to speak, is such a transformation of the creature's sensorimotor habitat that gives rise to the first genuine cognitive niche medium. Further on I will speak of a theoretically postulated First Cognizer, which is, of course, a fictional character, the ideal end of the downward ladder; nevertheless, it can be a heuristically useful device. For our goal is to understand how it comes about, in principle, that the cognitive niche medium is brought forth on the basis of a sensorimotor habitat.

Cognitive niche construction targeted as a series of transformations of the sensorimotor habitat (the base) can also be unpacked in functional terms, following the route taken by Di Paolo et al. (2017), but first of all following the functional characteristics of the niche spelled out by Hutchinson (1957). Doing so, we must only remember that there is a certain activity, real-life effort and work behind each of the simple mathematical posits. Keeping the latter in mind, we can now actually see the benefits of Hutchinson's abstract approach, for it enables us to provide a minimal, grossly simplified, but still useful characterization of the cognitive niche. Within this functional characterization we shall be able to distinguish objects, processes, and events, for these had to emerge in the surroundings of the fictional First Cognizer.

More to the point: we want to characterize a family of functions, let us label them *CN* (which stands for *C*ognitive *N*iche), describing the functioning of the First Cognizer in its sensorimotor habitat. Without further ado, I propose that all *CN*-functions shall be 3-argument functions involving the following variables:

First, the codomain (the set of destinations of all *CN*-functions) consists of sets of invariants $inv_1,\ldots,inv_n$, generated in the sensorimotor habitat; crucially—not of invariants themselves, but already certain clusters of invariants attributable to a fixed habitat or part thereof.

The next step is crucial, for it introduces a genuinely new element to the model built by Di Paolo and collaborators. I propose that the domain of *CN* (the set of departures) consist of two sets of variables. One is by no means controversial—we have time or rather a fixed measure of time decomposed into a finite set of points/moments $t_1, \ldots t_m^2$. This is the time-frame determined by the cognizer. But the other set is composed of what I propose to call, following Smith and Brogaard (2002) *granular partitions* (in short, I'll speak of partitions—*p*). This is meant to introduce the neglected mereotopological dimension to the model, that is, a dimension intended to do justice to the already discussed fact that the entities that we all actually deal with *as cognizers* are usually bounded clusters of parts with an address. The partition variable *p* is supposed to cover this aspect, while there is no room for a proper mereotopological treatment. As announced, this will be our *ersatz* ontology.

We can read first:

> A granular partition is a way of dividing up the world, or some portion of the world, by means of cells. (Smith & Brogaard, 2002, p. 157)

By "cells" they mean bounded units that "divide up" the world as if cutting it into pieces:

> We are to think of a granular partition—sometimes referred to simply as a 'partition' in what follows—as a grid of cells laid between ourselves and reality (or between ourselves and whatever is the relevant object domain) in such a way that its cells are transparent. These cells allow the objects in the domain over which the partition is laid to show through in undistorted fashion. Consider the way in which in constructing weather maps we may project a grid of political divisions onto satellite images. (ibid., pp. 155–156)

Here is the intuition behind the posit: think of any perceived scene. If you see a group of trees, you can say you see *a forest* and *a house* in front of it: the partition you impose draws a boundary between one being (one cluster if invariants) called "house" and the other being called "forest." So there is a system of partitions written into perception itself, such that you can do nothing about it (such partitions are responsible for the most fundamental "architecture" of percepts, e.g., figure-ground distinctions), but there is also a level you can manipulate to the effect that you change the partition imposed by your attention and, for example, begin to see individual *trees*; and you see that the house is an example of late modernism due to the lack of any decorative elements and others. Or take a different example—one that we shall come back to later on, yet based on the example already invoked in Chap. 3 (which should give us some initial sense of what the final posit of the chapter will be). Namely think of Oswald as he is shooting Kennedy. In step with one partition, you can take his pulling the trigger as a process (from a different angle—as an act), but note that "pulling the trigger" is different from "the bullet causing damage to Kennedy's body." It may have turned out that Oswald did not want to kill, but just to scare everyone; and it may have turned out that it was not Oswald's shot that killed Kennedy but the second shooter's, who became a sort of mythical figure in conspiracy theories. And most importantly, pulling the trigger is not a crime while killing usually is; and the bullet causing damage to Kennedy's body does not represent a crime either, especially as a bullet cannot be charged with anything. So the partition adopted here is somewhat strange, but thinkable and perhaps even useful if we want to look at the tragic event from a more unusual angle (let us keep this in mind!). On the other hand, the standard partition of the discussed realm of invariants is represented by something like "Oswald killing Kennedy," which clearly refers to a crime.

Smith and Brogard write that each partition has a certain resolution. I would say instead that the resolution of a certain manifestation of the world can be spelled out *in terms of* granular partitions: depending on how we cut the world into cells, we can see it as more or less fine-grained or granulated.

Two aspects or two theoretical posits are merged in the notion of granular partition: we have the concept of boundary and the idea that by partitioning the world, a certain setup of boundaries between entities is drawn so that it becomes possible to speak of the world as having a certain structural stability. Then there is location—cells determine exact locations to the effect that once the world is partitioned, each bounded cell is taken as occupying its own, perfectly filled "hole" in the world (Smith clearly carries on his mereotopological project here, launched in Smith, 1996).

One thing, however, must be approached differently from how Smith and Brogaard (2002) originally devised it. They draw a clear distinction between outside reality and the partitions "objectively" existing in it, and various partitions imposed on it by the intellect. As part of biological externalism we must reject the mind vs. world dichotomy as well as their tacit intellectualism. First, the partitions that emerge over the course of cognizer-environment interactions are not fiats; they may be stable or manipulable to various degrees, but this has nothing to do with the mind vs. world divide. In short, various ways of carving the world into cells are outcomes of a living creature's dealing with its surrounding and they are as real as anything else. Secondly, cognizer-dependent partitions are not solely of the intellect's making; the worm which digs its cavity in moistened soil also "draws" certain partitions; a potter who creates a jug by altering the shape of clay manipulates certain partitions, since both boundaries and the exact location of the clay (the "hole" in the world it occupies) change over the course of this process. Therefore, we must think of partitions as something embodied in the world itself, and also as something enacted in the world by various bodily engagements. Finally, the *mind vs. world* divide is also a partition, which was not problematized by Smith and Brogaard; it's not there, pre-given; it also results from a certain system of cells.

So, p stands for a certain granular partition (or partitions) relative to the cognitive capacities of a given cognizer.

Once the latter posit is set, it is finally possible to characterize a family of *CN*-functions. As for their domain and codomain, we have:

(27) $CN: \{t, p\} \mapsto inv$

The basic intuition behind *CN* is that our fictional First Cognizer constructs its cognitive niche, thus transforms the niche base into a niche medium, by taking some segments of its sensorimotor habitat composed of fixed invariants, and providing it with a certain mereotopological order of partitions, unfolding in time. Prior to that "moment," the partitions were not so much left undetermined—rather, they were out of the question. What the First Cognizer does differently in the world when compared to non-cognitive agents is precisely this—*bringing forth* a certain more or less stable structure of partitions in its surroundings. Simply put, the intuition is that cognizers are *boundary-drawers*, metaphorically speaking. This does not mean, though, that the cognizer picks a partition it likes out of the blue; the partitions—at least their basic level—are determined by the conditions of autonomy and by the nature and sensitivity of the cognizer's cognitive capacities. Once this basic level of partitions is determined, it at last *begins to make sense* to talk of less determined partitions or even undetermined ones. This indeterminacy is best exemplified by the differences between various conceptualizations and levels thereof (beginning with the basic level categorization which is rather fixed, hardly flexible), yet keeping in mind that the latter applies only to relatively sophisticated cognizers (perhaps only humans). Attentional mechanisms and their shifts can also be interpreted as partition shifters and they are more widespread in the animal kingdom.

Having said all that, though, it must be emphasized that partition flexibility is not part of our basic *CN* which is supposed to be true of the functioning of the fictional First Cognizer. The latter is too "primitive" for partition flexibility. This means that within any concrete exemplification of *CN*, partitions are taken as robustly fixed. However, as we shall see, partition flexibility is, in a sense, the birth mark of the capacity to problematize which builds upon the niche medium defined by *CN* (to be characterized shortly).

It is possible now, by adopting the method employed by Di Paolo et al. (2017), to define a new kind of invariant—this will be an invariant relative to *CN*. We can also make use of partial derivatives. Suppose we have the partial derivative function $\partial CN/\partial t$, assuming that p is constant. The latter assumption is in step with the above discussion of the flexibility of partitions and their relative rigidity within *CN*.

Therefore:

(28) When for some concrete exemplification of a *CN*-function we have $(\partial CN/\partial t) = 0$, relatively to a fixed time interval $t_n, \ldots t_i$, and provided that p is constant; and if the *CN*-function is continuous in that fixed time interval, then we can say that the *CN* represents an *object*—a being that lasts in time, at least in the fixed time-slice, without any change of invariants.

Simply put, objects are time-stable clusters of invariants, provided with a fixed, relatively stable partition. Further on:

(29) When $(\partial CN/\partial t) = a$ in the fixed interval, we can say that the CN-function represents a *process (provided that all other conditions from 28 apply).*

Processes are therefore beings that unfold in time in a continuous manner so that for each step from one moment in time to another we have a proportional change in the realm of invariants, also provided that partition *p* is fixed.

Finally:

(30) If a *CN*-function is not continuous in the given interval, then we can say that the function represents an event, or a series of events.

We can imagine an event as either a point-like occurrence, or a sequence of such occurrences with a clear beginning and end (thus as a fixed set of points). It's characterization is here, as part of this provisional framework, purely negative—events are not object and they are not processes, at least from the perspective of a fixed time-slice; they are occurrences that do not last and do not unfold in time, that's all we need to know at this point.

This is our *ersatz* ontology with objects, processes, and events. In this light, what distinguishes the fictional First Cognizer is the fact that the patterns of interactions it is capable of maintaining with its sensorimotor habitat *bring forth the first-ever system of partitions and partition-determined "architecture,"* locally distinguishing segments (partition cells or groups thereof) enduring in time (objects) or linearly changing in time (processes), thereby also making room for occurrences that do not fit this relatively robust and, in a sense, static structure—for point-like events.

Each cognitive niche medium, therefore, or the basic level of each cognitive niche construction could be given a lengthy description containing a variety, sometimes a huge variety, of concrete *CN*-functions referring to specific collections of invariants, their mereotopological scaffoldings provided by fixed partitions, their timeline, and a description of how the invariants behave in time (thereby indicating all the relevant objects, processes, and events). Note in this context a difference and at the same time an affinity between this simplified model of cognitive niche construction based on invariants and Hutchinson's (1957) original model of the ecological niche. Hutchinson targeted certain environmental conditions, primarily resources, which we have abandoned, but what is even more pregnant is the very manner in which he approached the concept of the niche; the manner adopted here for our own purposes.

Now, someone might point out, quite rightly, that all of these posits are theoretical fictions. We never encounter entities that do not change in time; it is just a matter of the limited resolution of our perceptual or intellectual grasp; processes are never perfectly linear, and so on. What we take for an object may turn out to be a combination of processes, a sudden event may turn out to be part of a process we just did not know of, and so on. This is all true, of course, but in order to discover these more fine-grained dynamics, a cognizer must be able to problematize its surroundings, to manipulate partitions to some extent, and, as I intend to show, the theoretical fictions just unpacked help in the task of understanding how the procedure of problematization works. So bear with me and with them for a moment.

All that being settled, we can re-introduce the information talk adopted by Pinker and others. As pointed out, the main problem with the informational approach to the cognitive niche was that it took information as if it was a kind of raw material, a resource waiting to be extracted and exploited, like coal or gas. Let us then try a different approach.

To begin, it is clear that only events can actually serve as vehicles of information, for if the rate of change of an object in a limited time interval is zero, then at least the "subjective" probability of the subsequent occurrence of an invariant from the relevant series—the "subjective" probability attributable to the object, to put it less aptly but more intuitively—is so high (whether the probability is 1 is debatable, I leave it for now) that one can hardly say that *uncertainty* has been reduced. Objects are entities we are quite *certain* about. The talk of "subjective" probability is justified insofar as we take information as something to be received and exploited by an actual cognizer (see Floridi, 2006, 2011), not as an "objective" measure of the complexity of a certain structure. The same applies to linear processes; they are "subjectively" certain, too.

Therefore, the amount of information attributable to objects and processes is actually zero or close to zero, and the only genuine vehicles of information are events whose "subjective" probability, again limited to a given time interval and fixed partition, may vary. Yet, crucially, events can only stand out as less predictable occurrences against the background of objects and processes. Hence, objects and processes carrying zero information set up an environment for more (potentially) information-loaded entities; they make the flow of information, as Dretske (1981) put it, possible while not being part of the flow themselves.

The cognitive niche medium can therefore be targeted as a realm of information in the sense that it is a relatively stable structure made of sensorimotor invariants provided with a mereotopological structure, which enables cognizers to take some events as vehicles of information.

In sum, the cognitive niche medium can be represented as a family of functions of the *CN* type; some of these functions *behave as* constant functions and some *behave as* linear functions within the limits of fixed time-slices, provided that partitions are constant; and finally some of these *CN*-functions resemble neither constant nor linear functions. In the latter case the relevant invariants are potential vehicles of information.

5.12 From Cognitive Niches to Problem Spaces: An Introduction to Embodied Erotetics

We now have a provisional grasp of the cognitive niche base and of the cognitive niche medium. It is only now that we can say more on what actual *construction* of a cognitive niche is, insofar as it means not only the first transition, so to speak, from the level of sensorimotor habitat, thus the base, to the proper cognitive niche medium, but an actual *use* of the latter.

There is likely a variety of factors that prompt cognizers to actively manipulate and transform their surroundings, leading to complex niche engineering such as what the already mentioned nest-building birds or beavers are capable of, not to mention humans. However, I want to make precise the necessary condition that our fictional First Cognizer has to satisfy in order to be able to take this path in the first place. The partitioned niche medium must be somehow brought into action by the First Cognizer; partitioned cells must serve as affordances in Gibson's (1979) sense and for that sake, I argue, *they must be problematized*. For clearly not all objects and processes can play the role of affordances; otherwise, the notion of affordance would become inflated (if everything is an affordance, then nothing in particular plays that role and the very talk of affordances becomes futile). Now the question is how it comes about that some ingredients of the relatively stable architecture of the cognizer's niche medium start to serve as potential tools or potential challenges.

One answer quickly comes to mind: genuine cognizers, thus at least organisms that can be regarded as cognizers without the shadow of a doubt, manifest curiosity and *exploratory behaviors*. They are active seekers of information and therefore have the capacity to spontaneously learn more about their cognitive niche medium.

As, for example, Kidd and Hayden (2015) note, it is a challenging task to elicit and do justice to all theories devoted to curiosity, for sometimes the term itself is not used by the researchers studying the phenomenon in question. Moreover, there is no single, universally accepted definition, and in part due to that lack, curiosity studies are mixed with investigations into animals' dealing with novelty, play, exploration, and so on, all of which clearly have to do with curiosity.

Problematization is not, or—to put more cautiously—may not be identical to curiosity, depending on how we understand the latter, and for that reason I would like to clarify the conceptual framework of this book so that my posits are clear and open to criticism for what they really stand for, not for what they allegedly inherit from their conceptual cousins.

A major intricacy having to do with the concept of curiosity, and thus a major reason why I put forth a new term—problematization (not a new word, of course, but a new technical posit)—is that it is not clear what curiosity has to do with novelty. For example, we can read that "[r]egardless of the specific theoretical claims, the key element in curiosity-stimulating mechanisms is the novelty of the stimulus" (Pisula, forthcoming). Meanwhile, the point is that problematization, as we shall see, is *not* a reaction to novel, unknown, and uncertain occurrences that attract our cognitive tools like magnets. On the contrary, its specificity boils down to the idea that in addition to dealing with novelty, we need a cognitive capacity to *deal with the familiar*, so that it doesn't stifle our cognitive abilities. In addition to dealing with uncertainty, we need a capacity to *deal with too much certainty* so that we are not deceived by it.

Loewenstein (1994) says that curiosity is the need for new information, and in order to strive for more information one must realize that the information one currently has, referring to a given entity, is not everything there is to be known about that entity (see also Litman et al., 2005; Inan, 2012; Misčevic, 2020). In short, one recognizes a lack on one's informational account, so to speak, and then strives to fill it. What does this have to do with problematization? Someone could argue that the latter works exactly the same way, namely when we problematize, we question phenomena that we want to learn more about. However, there is one subtle thing that I would like emphasize—one has to recognize that one is short of information, and insofar as objects and processes are concerned, this is impossible in principle, at least within the *CN* account rolled out above. For objects and processes by their nature convey zero information, so without an additional factor that could intervene, there can be no need to get more information, thus to learn more about them; they don't force themselves on us. On the contrary, when a novel event occurs, it may be a vehicle of information since the cognizer is initially uncertain about what happened and thus can naturally demand more information. Curiosity can therefore be seen as a mechanism responsible for filling the newly emerging information deficit. This phenomenon has been investigated in detail.

To get a sense of what the pursuit of problematization is about, let me bring in the notion of sense-making which enactivists emphasize. If we distinguish active sense-making and its relatively robust products—the meaningful, familiar world that we experience—then the products of sense-making composing the medium of the cognitive niche no longer take part in the process of active sense-making. Instead, they constitute a realm in which the pursuit of active sense-making can be realized, which is, of course, a critical thing, yet it is nevertheless different from actual partaking in the process itself. Think of navigation and familiar orientation points along the way, such as recognizable buildings. When someone says "turn right once you pass the cathedral" the cathedral does play a role in my sense-making activity at that moment, but not in the same way as, for example, the newly launched construction works that force me to cross the street and then circle around them in order to go back to my planned route, and so on. The cathedral is *not an issue* for me at that moment, however spectacular it may be for a tourist or a painter; it serves as mere background for my active sense-making. Speaking of painters, suppose that the mentioned cathedral is in Rouen and think how different the situation must have been for Claude Monet when he was making his careful studies of the light influencing the surface of the building. What happened as a result of Monet's act of problematization can be summarized as follows: a certain object, a relatively robust product of sense-making, conveying no information, was *re-introduced into active* sense-making—the process of observation and painting in this case—and as a result became, or some aspects of it became, vehicles of information. Long story short, an informationally futile entity was transformed into an informationally fecund one. This is the "miracle" we owe to acts of problematization. Yet the latter must now be simplified and, in some sense, operationalized in the functional terms adopted following Di Paolo et al. (2017).

Recall that when objects and processes were provisionally defined within our functional *ersatz* ontology, the mereotopological variable p was taken as constant. We were interested in how clusters of invariants within a given *CN*-function change in time, assuming that the mereotopological framework of partitions imposed on the invariants is fixed. Meanwhile, the key point of my idea about problematization is making p into a genuine variable. So the first intuitive approximation says that when we problematize something within our cognitive niche (within its medium, to be strict), we take time as constant and focus on how clusters of invariants change, if they do, in connection with the manipulation of partitions. So this time we have the partial derivative $\partial CN/\partial p$. In other words, when we problematize a given entity, the question is how the combinations of invariants generated in our sensorimotor habitat would change if we tried out *different* partitions.

Remember that there must be some real-life activities behind all these simple mathematical posits. So changing partitions with the assumption that time is constant requites some specific acts, yet not only acts of imagination, as someone might think, but also manipulative acts. Remember that unlike Smith and Brogaard (2002) we decided to take partitions as embodied; thus, manipulating partitions can also be taken as a genuine bodily activity. For example, when a child takes a plastic rectangular brick and tries to put it into a circular hole, time is irrelevant and the only thing that matters is whether the invariants composing the brick change in accordance with the desired partition—one that would fit the circular hole. It's obvious for us, more or less skilled problematizers, that this particular setup of invariants is immune to most of our manipulative acts, but the child must figure it out. That said, if the thing was made of plasticine, the situation would have been much different, for the sensorimotor invariants would have been much more flexible. Of course, the pursuit of altering partitions also involves highly sophisticated acts of the intellect and imagination; yet, the point is that at the bottom of this complex variety, there is some theoretically presumable "most primitive" capacity to bring forth and then manipulate a mereotopological "architecture" of partitions; to see the world in cells and then to try to juxtapose and alter these cells.

However, as pointed out above, the peculiar role of problematization is not to derive more information from potentially information-pregnant events. Instead, it is a very special capacity to activate some information-pregnant aspect of objects and processes that normally do not convey any information. We therefore have to limit our attention to, and broaden our understanding of, functions that describe objects and processes. In other words, sudden events call for scrutiny in and of themselves, for example, the event of killing JFK called for an investigation; thus, it *already* manifested itself *as a problem* (given the context and the cognitive capacities of human beings, etc.); yet objects and processes need to be *brought into presentation* (to use Husserl's great phrase again) *qua* problems.

Here is a semi-formal account. First, suppose for the sake of this presentation that we have a concrete object labeled here as # and the *CN*-function describing the cognitive niche medium in which # is enacted: *CN#*.

I propose that an act of problematizing # starts from a pursuit of finding the partial derivative of *CN#* with respect to p, assuming that time is constant, which means checking in what way and to what extent the invariants constituting # change once the partitions change; crucially—partition change must also be limited to a fixed range called the *range of partition manipulation*.

The entire time interval $t_n,\ldots,t_m$ attributable to #, thus the time-slice that # is relative to, can be "represented" here, for the sake of simplicity, by one element of the set; let us assume that it is $t_{\#}$. For, as mentioned, while time is a necessary dimension if we want to say what it means for any object to be *an* object, actual time flow is irrelevant when it comes to the pursuit of problematization. So when we refer to the partial derivative of *CN#* with respect to *p* we assume that time is constant; that it is always $t_{\#}$.

Let us also assume that $p_{\#}$ refers to the partition attributable to #; the partition that constitutes our default object #. By definition it belongs to $p_i,\ldots p_j$—as the partition that every manipulation departs from, it obviously belongs to the range of possible partition manipulation.

Finally, let *PR*( ) mean a function of problematization. I propose that:

$$(31) \quad PR(\#) = \frac{\partial CN\#}{\partial p},$$

relatively to the fixed range of partition manipulation $p_i,\ldots p_j$.

Simply put, the problematization of # can be represented as the inquiry into how the function *CN#* behaves within the range $p_i,\ldots p_j$. Crucially, the inquiry into the behavior of *CN#* is not part of *CN#*; it is clearly a new function, which means that problematization *builds on* a specific cognitive niche medium, but it is, in a sense, the next step in the development of the niche. In other words, the capacity to problematize is a genuinely new capacity and it is likely that many relatively "primitive" cognitive niche builders do not have it.

Now, what can we learn from the inquiry into the behavior of *CN#* within the fixed range $p_i,\ldots p_j$,? First, it may be the case that nothing changes.

This would mean, basically, that # is *totally immune to partition manipulation.*

Let us keep in mind this immunity, for philosophy shall be especially interested in it; however, a more realistic situation is when the two sets of invariants are not identical. Therefore, we can have a situation in which there is a difference between the invariants constituting # and the ones obtained by various manipulations of the partition originally attributed to #.

We can name this difference the *outcome* of the problematization of #. What we learn from it is how much *CN#* deviates from the invariants constituting # once the partition attributable to # gets changed within a certain manipulation range; in other words how partition-sensitive the object # is.

From a more practical angle, if we discovered that # was totally immune (if possible) to partition manipulation, what that discovery would mean is essentially this: there is nothing to learn from # in the sense that # is absolutely informationally inert. Yet this most likely rarely happens, so the non-empty outcome of problematization shows how information-pregnant # *potentially* is.

Before we move on, let us consider an example. Monet's case is too complex to fit the primitive scheme rolled out above, and it also involves time as an independent variable, but on the other hand it is appealing. We can therefore simplify it and try to identify the *initial act* of problematization that Monet could have exercised before he started to manipulate time. For time manipulation was already part of the process of painting which was not an act of problematizing any more, but something spurred by the problematization he executed. But what was the latter about? Suppose the cathedral as it appears to non-painters and initially also to Monet was a fixed object, thus a cluster of partitioned invariants that did not change, including the color invariant—it was seen as simply being gray. In other words, suppose that the partition was so coarse-grained that nobody ever paid any attention to the fact that the color manifested at the cathedral's surface was not just gray, but a more complex cluster of shades relative to lighting conditions (the assumption is hardly realistic, but let's make it here provisionally). In ordinary people's eyes the cathedral was simply *there*, a bulky gray object, basically intact. Now, what Monet initially did can be described as an act of problematization insofar as he tried to manipulate the initial coarse-grained partition or resolution a little bit within a fixed moment, and suddenly the previously stable color invariant "gray" was torn apart. With a different partition/resolution Monet was able to spot multiple color invariants, some of which—as he reflected on them further—must have been tied strictly to the fixed time-moment to which he limited his analysis. Later on, of course, he devoted himself to further observation of how these invariants changed over time, but this was, exactly, a process of observation unfolding in time, whereas what spurred it in the first place was not something that played out it time; instead, I would argue, it was *an instantaneous* act of problematization, that is, a manipulation of the initial coarse-grained partition.

Generally, it is crucial to draw a line between problematization and what comes afterward, for example careful observation, experimenting, as well as what psychologists would call curiosity as an active search for information. For all these activities are different from, while at the same time being provoked by, what I am calling an *act* of problematization, functionally characterized in the simplistic way above. In this context, speaking of the *pursuit* of problematization, which I do in this chapter, is somewhat less technical and refers to a more complex phenomenon such as a certain pattern composed of acts of problematization, an intellectual habit or skill, or even a life-path.

It is somewhat strange that we are illustrating the model designed for the fictional First Cognizer, arguably a very simple theoretical posit, with the example of what is likely one of the most sophisticated instances of problematization—one made by a human of extraordinary sensitivity like Monet. Yet while the former is done for the sake of a theory, which must start from simple thinkable cases, the latter is intended to propel our imagination, to give us a more intuitive sense of what is being theoretically targeted here.

5.13 Philosophical Problematization

We can finally take on one of the most sophisticated and peculiar cases of problematization—the type of problematization pursued by philosophers. However strange it may sound at first, (although I am somewhat encouraged by Rowlands & Monsó 2017) we wish to describe what a fictional First Philosopher, thus our First Cognizer a couple of steps of cognitive development later, does. At the same time, to get a sense of what is at stake, we still consider the other extreme—not with Monet or any other famous individual this time, but *ourselves qua actual philosophers* (academic or not), trying to see our endeavors in light of the simplified model being put forward here.

In the previous chapter I argued that philosophizing begins with the recognition of one's epistemic position, which means becoming aware that the world we are *in* is *a* presentation. This is not to say that we live in a dream dreamed by our brains, for the distinction between dreams and reality is also part of the presentation in question, yet this suggests that there may be a certain plurality of manifestations attributable to a supposed reality—in itself unknown. Kant is likely the most recognizable figure in this context, but Plato's metaphor of the cave is equally well-known; both are basically about the same thing, about the same recognition.

What is special about philosophy is an attempt to identify what it means for the presentation we actually face, and any other possible presentation, to be *a presentation* in the first place; what is the relationship between a presentation and what presents itself; what is the most fundamental "architecture" of *our* presentations and of *all* possible presentations as such, thus what are the structural ingredients built into every manifestation of reality. In academic practice, as we pointed out, this boils down, in most cases, to investigating the general categories organizing a given realm—ontological categories which are the inventory of the broadest and most abstract domain of *what there is*; epistemological categories which are the "architectural" determiners of the domain of cognition; ethical categories which are the "architectural" determiners of the domain of interactions with other agents (human and non-human).

Thanks to the simple model rolled our above, we are able to spot an interesting difference between the problematizations exercised by the philosopher and those exercises in various other disciplines, especially the sciences. Recall that according to the proposal put forward in this chapter, problematization is a means of extracting information from what seemed to convey zero information within a limited time frame, thus of re-introducing relatively stable products of sense-making into the active sense-making process. Now, in this context it seems that philosophy is more concerned with what remains stable than with what changes. Beginning with Parmenides and Plato, philosophy has been quite explicit about the distinction between *becoming* and *being*. Putting it differently, the search for an ultimate system of categories, as the search for Platonic ideas, but also the Kantian/Husserlian search for the "architectural," transcendentally valid features of all presentations of reality—these are all *pursuits of stability* rather than change, purposely directed toward entities that convey zero information. For that reason, let us define a new entity which I propose to provisionally refer to as constant which is represented by all CN-functions relative to a given sensorimotor habitat such that $(\partial CN/\partial p) = 0$. This means clusters of invariants that do not change with the alteration of partitions—however we play with the partitions, these invariants stay constant—analogously to fixed time intervals—given a certain finite set of partition manipulations. The example of the child with the plastic or plasticine brick invoked earlier indicated how we may think of constants within clusters of sensorimotor invariants. Note that constants are different from objects: they are clusters of invariants that are insensitive to changes in partitions, whereas objects (in our ersatz ontology) are clusters of invariants that are insensitive to time flow.

It can now be said that philosophy is concerned with objects and constants. If so, one might claim that in order to succeed at these pursuits of zero-informational "architecutes," it is sufficient to simply stop problematizing and then everything will be zero-informational, period. Yes, but here is the point: I argue that philosophy aims at what is *genuinely* zero-informational, thus genuinely stable or constant, which requires, somewhat paradoxically, a never-ending process of painstakingly problematizing virtually everything, as if testing whether what seems constant and stable is genuinely constant and stable. In this sense, philosophers, as indicated many times in this book, problematize what is usually fully transparent. That said, there is also an interesting aspect to that somewhat negative advice to stop problematizing: perhaps the peculiar affinity between no problematization and extreme problematization, both seeming to result in the same outcome—an encounter with a variety of zero-informational entities—is responsible for two things: the apparent banality of philosophical claims, discussed in Chap. 4; but also, at a more profound level, the fact that while creating the most abstract vocabularies possible, philosophy also exhibits a tendency to purposely restore a naïve attitude to the world. Just think of such categories as Husserl's *Lebenswelt* or the relationship of being-in-the-world scrutinized over the course of this chapter in niche terms: they are designed as a refinement of naivety, which was nicely unpacked by Merleau-Ponty in his introduction to "Phenomenology of Perception" (Merleau-Ponty, 1962).

Philosophy is therefore *by design* a zero-informational cognitive endeavor, however peculiar this may sound. By the way, it explains why philosophy does not accumulate knowledge in the manner the sciences do.

Yet this specific concentration does not by itself make philosophy what it is, for it can be a part of other disciplines, too, however marginally, and of everyday life. What makes philosophy special, I argue, is the attitude with which this pursuit of objects and constants is undertaken; the attitude that philosophy owes to the recognition of one's epistemic position. This peculiar act, characterized in Chap. 4, must be targeted as a philosophy-specific act of problematization which provides the foundation for all other acts of problematization attributable to philosophy. The question is how this specific act of philosophical problematization can be represented—grossly simplified but still, somehow done justice to—by the model just rolled out.

First, note that while the model is simple, it is also very general. For there is no a priori restriction as to what can and what cannot be captured in terms of cognitive niche construction. In particular, we can speak of many kinds of invariants that serve as the actual building blocks of cognitive niche media at different levels of complexity and cognitive sophistication. These include not only sensorimotor invariants, but a much broader pool including basic invariants in our sensorimotor habitats, but also more complex perceptual invariants, conceptual invariants, emotional invariants; invariants of various behaviors; invariants of thinking; even invariants of doing science which encompass formal/conceptual procedures on the one hand and various patterns of how people act in laboratories on the other.

Based on this unrestricted approach to cognitive niche construction, there is also no a priori (pre-given) limit as to what can and what cannot be problematized. Intuitively speaking, therefore, philosophical problematization is not directed toward any specific domain of invariants, but instead problematizes the very pursuit of cognitive niche construction qua cognitive niche construction. In niche terms we can say that the recognition of my epistemic position means problematization of the "fact" that I *do* live in a cognitive niche—that I *am* in the Platonic cave; that I *am placed* and surrounded by my Umwelt. Further on, that this Umwelt or cognitive niche is species specific, perhaps to some extent individual (here the puzzle of qualia reemerges) and perhaps also transcendentally specific, which would mean, as Kant and Husserl argued, that there are certain conditions, certain "architectural" features that any Umwelt or any cognitive niche (any realm of phenomena) *must* have built in.

There is no reason to insist on there being just one privileged strategy leading to the recognition of one's epistemic position so understood. However, for the sake of this investigation, I propose that the recognition of one's epistemic position can be seen in the first place as an act of problematizing the very pursuit of problematization which, indirectly, leads to the problematization of all cognitive acts as such.

This loopy, self-referential act of problematization, directed *at itself* like an ouroboros eating its own tail, is a strategy that the First Cognizer living within our simplified model *may* adopt to recognize its epistemic position, thereby entering the domain of philosophical inquiry. But how might this ouroboros-like problematization actually work?

Suppose the First Cognizer problematizes a certain object in the world, say a concrete apple, and let the said problematization be represented here as the function *PR*(*apple*). Now, let's assume that the First Cognizer is clever enough—capable of what psychologists refer to as metacognition—to eat its cognitive tail in the ouroboros-like manner, as it were. Hence the First Cognizer strives to problematize *PR*(*apple*), as if asking himself "How did I manage to perform this problematization?" or even "Could I do it differently?" We thus have something like *PR*(*PR*(*apple*)). Note that *PR*(*PR*(*apple*)) differs from *PR*(*apple*), for what is being problematized now is not a thing in the world, but a certain cognitive activity. Here we will have some invariants of thinking instead of perceptual invariants, relevant objects, thus clusters of invariants of thinking that do not change in time, and so on. These invariants of thinking could include various categories and modes of reasoning, at least when it comes to humans, but imagine that the First Cognizer's cognitive acts also involve some invariants of this kind. In sum, the pursuit of problematization, *once itself problematized*, turns out to be a constrained endeavor just like the apple which is a perceptually constrained entity. The next thing the First Cognizer can do, although it involves a different cognitive capacity, is to replace "*PR*(*apple*)" in "*PR*(*PR*(*apple*))" by the sum of all the problematizations that our fictional character is able to pursue. Thanks to this maneuver, the First Cognizer may now recognize *all* the constraints inherent in *every* act of problematization.

Now, since an act of problematization is, as I argued earlier, built into every construction process producing a more complex cognitive niche—one that can be deemed a realm of affordances and transformed in the way described by Tooby and De Vore, Pinker, and Clark—the recognition just made by the First Cognizer automatically applies to everything the First Cognizer can cognitively achieve, so to speak.

Indirectly, therefore, problematizing the capacity of problematization equals the problematization of each and every act of constructing a more complex cognitive niche, in a single leap of the intellect. That's why it is such a surefire way for the First Cognizer to become the First Philosopher, of course within the limits of our simplified model. By problematizing its own capacity to problematize, the First Cognizer climbs up to an entirely new level of generality discussed in Chap. 4; actually, this new level of generality is brought forth by the ouroboros-like problematization of problematization.[3]

Now, as we pointed out, philosophers are more interested in what is constant than in what changes; as said, the crucial difference between philosophy and various other disciplines is that it is more interested in zero-informational entities than in information-pregnant ones. The question now is, what is the impact of the recognition of one's epistemic position on this search for stability? What kind of new zero-informational endeavors, so to speak, open up thanks to the recognition of one's epistemic position?

5.14 The Struggle for Marginal Transgressions: Philosophy as Form and Training

To answer the latter question, recall the following fact: each cognitive niche, whose medium is composed of objects, processes, events, and likely some other types of entities, is relativized to a certain time range.

Therefore, each act of problematization inherits this time-limitedness which is additionally suited to a certain range of partition manipulations. Objects, to take the most appealing case, were once considered clusters of invariants which remain unchanged over a period of time, but the time interval was always assumed to be fixed, even if effectively unknown to the cognizer. In the same way, the capacity to problematize is useful insofar as it can manipulate the partitions and thus present the object in a new light or resolution, thereby eliciting what is in fact changeable in it, but this manipulation only makes sense within a certain realistic range of variation. It is something like decomposing an entity that initially conveyed zero information into potential vehicles of information on the one hand, and their zero-informational "scaffolding," so to speak, on the other. Recall Monet's paintings. There was the cathedral-shape manifested on each picture, yet all the actual colors were changeable, hardly predictable, thus abundant in information. Noticing all this was possible thanks to Monet's ability to problematize the surface of the building.

However, our First Cognizer, thanks to the recognition of its epistemic position, thus thanks to the recognition of, among other things, these very limitations imposed on every single problematization, can at least "daydream," so to speak, about freeing itself from these limitations. Putting it more technically, the First Cognizer can therefore postulate that there are *objects unlimited by time-slices* and *constants unlimited by any specific range of partition manipulation*. These would be clusters of sensorimotor invariants that remain stable over an infinite time period, thus, actually, invariants that are stable independently of time whatsoever. Analogously, there would be clusters of invariants that remain stable over all possible manipulations of partitions, thus ones that would be fully independent of any partition whatsoever.

Hence, within our simplified model, the *pursuit of philosophizing*, if it is to be unpacked in functional terms, aims to take as its domain infinite time and the sum all possible partitions. It then seeks the codomain composed of all possible invariants. Finally, in this realm, it designs a function that might be represented as the sum of all possible specific CN-functions; let us label the latter as PCN. Philosophy, then, seeks to discover how PCN behaves, which means estimating the total derivative of PCN. By doing so, philosophy aims at something that we may deem *fundamental objects* (if any) thus objects that remain stable independently of time, and *fundamental constants* (if any) which remain stable in all possible manipulations of partitions. These two groups of fundamentals are supposed to determine the entire realm of the cognizable; traditionally speaking—they determine the fundamental "architecture" of *the intelligible* as such. Think in this light of projects like Plato's theory of ideas, or the project of eliciting the transcendental properties of all beings *qua beings*, but this also applies to non-metaphysical or openly anti-metaphysical projects such as Kant's or Husserl's transcendental analyses, or even, to some extent, to Carnap's Aufbau. It applies to the traditional projects of epistemology dedicated to unveiling what it means in principle—thus with reference to *any* domain of cognition, realized by humans or any other intelligent creatures (populating, say, one of Stanisław Lem's novels)—to *know* something, to say something *true*, and so on.

This ambitious pursuit is, of course, vulnerable to the criticism already invoked in Chap. 4, namely that, first, it is in principle a non-starter since we cannot free ourselves from the limitations that usually constrain our cognitive niches; and second, there can be no strategy of verifying such philosophical claims, for there is nothing they can be compared to, confronted with, or in any other manner set against the background of. Therefore, pursuing such ambitious projects—immodest, as Machery (2017) might say—should be abandoned. They aim, to cite Machery again, far beyond the limits of our epistemic reach.[4]

My answer to these concerns is also familiar to the reader of Chap. 4, yet here we can see it with more clarity perhaps due to the simplicity of the model. The argument from epistemic limits, in my reading, *assumes* that a range of doable partition manipulation is *already set* or pre-given, and the only problem with overambitious philosophers pursuing immodest, over-the-top projects is that they don't want to accept this limitation or even notice it.[5] Now, it is certainly a reasonable point, perhaps even a trivial one, that there *are* epistemic limits for each cognitive endeavor. This is basically one of the leitmotifs of this book. However, at the same time, it is self-refuting to refer to these limits as if they set up a fixed, insurmountable barrier to the ambitions of philosophy. For this would mean that while the traditional (philosophical) pursuits of fundamental objects and fundamental constants are in vain—which is what the argument from epistemic limits purports to prove—if the proclaimed reason for their being in vain is that there *are* fundamental limits, the latter somewhat paradoxically, can themselves be seen as immune to time and immune to all manipulation. Thus these limits themselves turn out to be fundamental objects or fundamental constants. In other words, the argument from epistemic limits, which is itself a product of philosophizing, has to make one concession, namely that philosophy, as constrained as it is, can nevertheless reach to one (and only one) fundamental, unchangeable (thus ideal in the Platonic sense) aspect of reality: the one that has to do with limits of reasonable philosophical projects. This paradox, rendering the argument from epistemic limits itself overambitious, does not prove the argument plainly wrong; we just need to keep in mind that it is as constrained as anything else.

Therefore, if there is no ready-made, pre-given range of available partition manipulations, the limits at stake *need to be actually tested.* A relatively modest way of doing this would require distinguishing between various types of cognitive endeavors, ordering them, and trying to seek objects and constants that transcend the *actual* limitations by a reasonable margin; let us call this process *marginal transgression* of the relevant limits as distinguished from the overambitious project of transgressing all the limits.[6] The term "transgression" is borrowed from the social sciences where it refers to breaching the limits of current social norms (see Jenks, 2003). It does not have this somewhat liberating or even revolutionary flavor here, but it appealingly unpacks the main idea: testing the limits of what *can be said and known* at the present stage of conceptual development. At the same time, the term "marginal" is taken from economics, where it relates to how a given variable, such as productivity, changes if another variable is altered within a fixed (rather small), measurable range; for example, if a factory hires one more worker marginal productivity increases, decreases, or remains the same. These two borrowed notions unpack the idea of what the pursuit of philosophizing aims at: breaching conceptual boundaries (transgression), while maintaining self-restraint (marginal).

This is basically what I meant in Chap. 4 by the pursuit of *boundary questions*. It is testing the capacities of the actual conceptual frameworks of specific disciplines by trying to transcend them by a *reasonable* margin.

Of course, what is reasonable is context-dependent and not determined in advance.

As for examples of marginal transgression that does not "go big" while retaining some of the traditional philosophical ambition of testing the limits of our epistemic reach, the best examples come from applied philosophies. Consider the debate regarding the definition of life in the philosophy of biology. Aside from the need for a definition covering all the living creatures known so far, one of the explicitly articulated demands which sparks controversies on its own is that the proper definition should also cover artificial living systems which are likely to emerge in the near future as well as living creatures populating other planets, for example, those that probably lived in the past on Mars, whose remains we may someday find. Clearly, the concept of life so understood goes beyond the actual limits of our cognitive reach by playing both with time intervals (when exactly should we expect any encounters with extraterrestrial life—nobody knows) and with partitions (in this case—with the range of applicability of the expected concept of life). In particular, this pondering of the nature of living systems transcends the limits of current biology, astrobiology, or AI, but—as someone might argue—it is still a reasonable transgression.

Take an example from a different field—all normative notions, such as justice, clearly go beyond the actual timetable by imposing a certain ideal and beyond the actual reference, meaning that even if there are actual entities that comply with the norm, the concept cannot just describe these entities (which would make it a descriptive notion instead); there must be more in it than what is attributable to these actual instances. If it makes sense to ponder the meaning of justice and the conditions that a just society must satisfy, then even if there *is* a thinkable society that can be called just, say the society of Narnia under King Peter, there must be more to that concept than what is realized by the Narnians if the concept is to set up a norm or ideal. For otherwise it would not be a normative concept of justice, but a peculiar descriptive concept of Narnia-under-kind-Peter, hence not a concept, but solely a name for the prototype attributed to the word "justice." Of course, someone might argue that all concepts boil down to prototypes, and there may be convincing arguments for that, yet my goal here was merely to show that we can think of relatively modest philosophical endeavors that retain some of the traditional ambition. After all, normative concepts are not considered outlandish and immodest by design. Yet they *do* transcend the limits of our actual epistemic reach, while keeping it a marginal transgression in the sense proposed above, insofar as they do not transform into full-blown utopias.

Finally, does it all mean that the traditional philosophical struggles to reach the unconditioned realm of ideas, fundamental objects, and fundamental constants are all non-starters? Should we conclude that they are all mere byproducts of the pursuit of problematization, or perhaps even some sort of overuse of the otherwise necessary pursuit of problematization and marginal transgression, in which the two reasonable intellectual efforts become mere art for art's sake? As for byproducts, I shall address this idea in the Epilogue. Meanwhile, I would like to devote a couple of final words to the latter point, that is, the notion of traditional philosophizing as art for art's sake.

I suspect that my answer will not satisfy many of those who would like to leave traditional philosophy be and defend its pretenses, but I do believe that there is a Socratic aspect to what I am about to say. The central concept in my reply to the critics of metaphysics or armchair epistemology (aside from what I want to put forward in the Epilogue) is *training*. I don't mean push-ups or spinning, but, insofar as sports are concerned, something closer to sequences of moves in karate, combinations of punches and kicks in the air, thus—as someone might mistakenly think—moves without a purpose; and finally the most artificial part of it, the so-called kata. All these sequences of moves are supposed to drill certain patterns of bodily movements and sensorimotor coordination, including patterns of instant reaction, so that they can be properly executed in a real fight or sparring. Had there been no painstaking hours of karate art for art's sake, as a critic might say, the fighter would be unable to act properly. Here is the analogy I submit to your consideration: it makes sense to go through, re-articulate and reinvent the traditional philosophical pursuits of ultimate and fundamental truths, since they are what philosophy is about *in pure form*, distilled from all actual uses and applications, just like karate sequences are pure fight forms distilled from real fights, yet they enable our cognitive faculties to acquire patterns of thinking that can be subsequently used for the sake of more specific, more modest marginal transgressions. It would be pointless to spar over the course of an entire karate training session, week after week and month after month, for one can learn little from that, perhaps contrary to what a layperson might think: you need to practice carefully selected sequences and then try them out in real sparing under strict conditions before you go big to a real karate tournament. By analogy, I argue that training with *pure metaphysical forms of problematizing*, or pure epistemological forms of problematizing, can only increase the sharpness of one's philosophical analysis in more specific contexts.

The training argument is Socratic in the sense that it arguably represents what Socrates himself was doing; and he was badly misunderstood. For his intention was not to overthrow the social order in Athens or confound official religion, but to teach people how to radically pursue problematization so that the pure form of problematization could sink in and be used in more modest instantiations when needed.

Notes

1. Not necessarily an animal, for there are attempts to see even unicellular organisms in pre-cognitive terms, as, for example, van Duijn et al. (2006) propose (see also Di Primio et al., 2000), not to mention the whole area of plant cognition; see for example, Garzon and Keijzer (2011); for a broader picture of cognitive evolution see Feinberg and Mallatt (2016).
2. This punctual measure of time is by no means uncontroversial, and most likely it would have to be rejected over the course of a more diligent phenomenology-based investigation (see Husserl, 1991; Ingarden, 1948/2013), yet I need a simplified account for the sake of the ersatz ontology being unpacked here. I thank the anonymous reviewer for drawing my attention to this issue.
3. Of course, many people may disagree with the presented account of philosophy, arguing that there are many possible accounts, while the one I put forward harks back to specific philosophical traditions not everyone adheres to. However, this the proposal I put on the table—I do believe that (keeping in mind how simplified the presented "model" is) it grasps some of the core features of the pursuit of philosophizing.

4. This also applies to Wittgenstein and even Carnap, yet their position here is certainly controversial, while, on the other hand, signifying a broader debate on the extent to which early analytic philosophy retained the bitterly criticized ambition of metaphysics.
5. The range of traceable time-slices is obviously limited by the fact that we all must die at some point. So, when philosophers seek fundamental objects, they must invent specific methods of analysis. My point is that it is unlikely that there are any pre-given limits to these methods.
6. By the way, Plato seems to have adopted this more modest strategy in his later writings.

References

Arfini, S., Bertolotti, T., & Magnani, L. (2019). Online Communities as Virtual Cognitive Niches. *Synthese, 196*, 377–397.

Bertolotti, T., & Magnani, L. (2017). Theoretical Considerations on Cognitive Niche Construction. *Synthese, 194*, 4757–4779.

Brancazio, N. (2020). Being Perceived and Being 'Seen': Interpersonal Affordances, Agency, and Selfhood. *Frontiers in Psychology*. https://doi.org/10.3389/fpsyg.2020.01750

Brooks, R. A. (1991). Intelligence Without Representation. *Artificial Intelligence, 47*(1–3), 139–159.

Brown, D. (2008). Descartes on True and False Ideas, in J. Broughton, J. Carriero (eds.) A Companion to Descartes (pp. 196–215) London: Blackwell

Casati, R., & Varzi, A. (1999). *Parts and Places. The Structures of Spatial Representation*. MIT Press.

Chemero, A. (2003). An Outline of a Theory of Affordances. *Ecological Psychology, 15*(2), 181–195.

Chemero, A. (2009). *Radical Embodied Cognitive Science*. The MIT Press.

Clark, A. (2005). Word, Niche and Super-Niche: How Language Makes Minds Matter More. *Theoria, 54*, 255–268.

Clark, A. (2006). Language, Embodiment, and the Cognitive Niche. *Trends in Cognitive Science, 10*(8), 370–374.

Cummins, F., & De Jesus, P. (2016). The Loneliness of the Enactive Cell: Towards a Bio-enactive Framework. *Adaptive Behavior*. https://doi.org/10.1177/1059712316644965

De Jaegher, H., & Di Paolo, E. (2007). Participatory Sense-Making: An Enactive Approach to Social Cognition. *Phenomenology and the Cognitive Sciences, 6*(4), 485–507.

De Jesus, P. (2015). Autopoietic Enactivism, Phenomenology and the Deep Continuity between Life and Mind. *Phenomenology and the Cognitive Sciences, 15*, 265–289.

Di Paolo, E., Buhrman, T., & Barandiaran, X. (2017). *Sensorimotor Life: An Enactive Proposal.* Oxford University Press.

Di Paolo, E. A. (2005). Autopoiesis, Adaptivity, Teleology, Agency. *Phenomenology and the Cognitive Sciences, 4*(4), 429–452.

Di Primio, F., Müller, B. S., & Lengeler, J. W. (2000). Minimal Cognition in Unicellular Organisms. In J. A. Meyer, A. Berthoz, D. Floreano, H. L. Roitblat, & S. W. Wilson (Eds.), *SAB2000 Proceedings, Supplement* (pp. 3–12).

Dretske, F. (1981). *Knowledge and the Flow of Information.* MIT Press.

Dretske, F. (1995). *Naturalizing Mind.* MIT Press.

Dreyfus, H. (1991). *Being-in-the-World.* MIT Press.

Dunbar, R. (2014). *Human Evolution.* Pelican Books.

Elton, C. S. (1927). *Animal Ecology.* Macmillan.

Feinberg, T. E., & Mallatt, J. M. (2016). *The Ancient Origins of Consciousness. How the Brain Created Experience.* MIT Press.

Floridi, L. (2006). The Logic of Being Informed. *Logique et Analyse, 49*(196), 433–460.

Floridi, L. (2011). *The Philosophy of Information.* Oxford University Press.

Garzon, P., & Keijzer, F. (2011). Plants: Adaptive Behavior, Root-Brains, and Minimal Cognition. *Adaptive Behavior, 19*(3), 155–171.

Gibson, J. J. (1979). *The Ecological Approach to Visual Perception.* Houghton Mifflin.

Godfrey-Smith, P. (1996). *Complexity and the Function of Mind in Nature.* Cambridge University Press.

Godfrey-Smith, P. (2016). Individuality, Subjectivity, and Minimal Cognition. *Biology and Philosophy, 31*(6), 775–796.

Gould, S. J., & Lewontin, R. (1979). The Spandrels of San Marco and the Panglossian Paradigm: A Critique of the Adaptationist Programme. *Proceedings of the Royal Society of London B, 205*(1161), 581–598.

Heidegger, M. (1962). *Being and Time.* SCM Press.

Heidegger, M. (1982). *The Basic Problems of Phenomenology.* Indiana University Press.

Husserl, E. (1913/1983). *Ideen zu einer reinen Phänomenologie und phänomenologischen Philosophie, I, Buch: Allgemeine Einführung in die reinen*

Phänomenologie. Max Niemeyer Verlag. Quotations from: E. Husserl, *Ideas Pertaining to a Pure Phenomenology and to a Phenomenological Philosophy. First Book: General Introduction to a Pure Phenomenology*. Martinus Nijhoff Publishers/Kluwer.

Husserl, E. (1991). *On the Phenomenology of the Consciousness of Internal Time*. Kluwer Academic Publishers.

Hutchinson, G. E. (1957). Concluding Remarks. *Cold Spring Harbor Symposia on Quantitative Biology, 22*(2), 415–427.

Hutto, D., & Myin, E. (2013). *Radicalizing Enactivism: Basic Minds Without Content*. MIT Press.

Inan, I. (2012). *The Philosophy of Curiosity*. Routledge.

Ingarden, R. (1948/2013). Spór o istnienie świata, Tom 1. PAU. English Translation: *The Controversy of the Existence of the World*. Peter Lang.

Jenks, C. (2003). *Transgression*. Routledge.

Keijzer, F., & Arnellos, A. (2017). The Animal Sensorimotor Organization: A Challenge for the Environmental Complexity Thesis. *Biology & Philosophy, 32*(3), 421–441.

Kendal, J. R., Tehrani, J. J., & Odling-Smee, J. (2011). Human Niche Construction in Interdisciplinary Focus. *Philosophical Transactions of the Royal Society B: Biological Sciences, 366*(1566), 785–792.

Kidd, C., & Hayden, B. Y. (2015). The Psychology and Neuroscience of Curiosity. *Neuron, 88*(3), 449–460. https://doi.org/10.1016/j.neuron.2015.09.010

Kirchhoff, M. D. (2016). Autopoiesis, Free Energy, and the Life–Mind Continuity Thesis. *Synthese*. https://doi.org/10.1007/s11229-016-1100-6

Laland, K. N., Odling-Smee, F. J., & Feldman, M. W. (1999). Evolutionary Consequences of Niche Construction and Their Implications for Ecology. *Proceedings of the National Academy of Sciences, 96*, 10242–10247.

Laland, K. N., Odling-Smee, F. J., & Feldman, M. W. (2000). Niche Construction, Biological Evolution and Cultural Change. *Behavioral and Brain Sciences, 23*(1), 131–175.

Laland, K. N., & Sterelny, K. (2006). Seven Reasons (Not) to Neglect Niche Construction. *Evolution, 60*, 1751–1762.

Leonard, H. S., & Goodman, N. (1940). The Calculus of Individuals and Its Uses. *Journal of Symbolic Logic, 5*, 45–55.

Leśniewski, S. (1916/1992). Podstawy ogólnej teoryi mnogości I, *Prace Polskiego Koła Naukowego w Moskwie, Sekcya matematyczno-przyrodnicza*, Moscow; English Version: Foundations of the General Theory of Sets I. In S. Leśniewski, *Collected Works*. Kluwer 1992, 129–173.

Lewontin, R. (1983). The Organism as the Subject and Object of Evolution. *Scientia, 118*, 63–82.
Lewontin, R. (2000). *The Triple Helix. Gene, Organism, and Environment*. Harvard University Press.
Litman, J., Hutchins, T., & Russon, R. (2005). Epistemic Curiosity, Feeling-of-knowing, and Exploratory Behaviour. *Cognition & Emotion, 19*(4), 559–583. https://doi.org/10.1080/02699930441000427
Loewenstein, G. (1994). The Psychology of Curiosity: A Review and Reintrepretation. *Psychological Bulletin, 116*(1), 75–98.
Machery, E. (2017). *Philosophy Within Its Proper Bounds*. Oxford University Press.
Maturana, H., & Varela, F. (1980). *Autopoiesis and Cognition. The Realization of the Living*. Kluwer.
Merleau-Ponty, M. (1945/1962). *Phénoménologie de la perception*. Available in English: *Phenomenology of Perception*, trans. by C. Smith. Routledge & Kegan Paul.
Miłkowski, M. (2017). Situatedness and Embodiment of Computational Systems. *Entropy, 19*(162). https://doi.org/10.3390/e19040162
Misčevic, M. (2020). *Curiosity as an Epistemic Virtue*. Palgrave Macmillan.
Moreno, A., Umerez, J., & Ibañez, J. (1997). Cognition and Life. The Autonomy of Cognition. *Brain and Cognition, 34*, 107–129.
Morgan, C. L. (1894). Introduction to comparative psychology, London: Routledge.
Noë, A. (2004). *Action in Perception*. MIT Press.
Noë, A. (2009). *Out of Our Heads. Why You Are Not Your Brain, and Other Lessons from the Biology of Consciousness*. Hill and Wang.
O'Regan, K., & Noë, A. (2001). A Sensorimotor Account of Vision and Visual Consciousness. *Behavioral and Brain Sciences, 24*(5), 939–973.
Odling-Smee F. J. (2010). *Niche inheritance*. In: M. Pigliucci, G. B. Müller (eds) Evolution: the extended synthesis (pp. 175–207), Cambridge MA: MIT Press.
Odling-Smee, F. J., Laland, K. N., & Feldman, M. W. (2003). *Niche Construction: The Neglected Process in Evolution*. Princeton University Press.
Piaget, J. (1969). *Biology and Knowledge*. University of Chicago Press.
Pinker, S. (2003). Language as an Adaptation to the Cognitive Niche. In M. H. Christiansen & S. Kirby (Eds.), *Language Evolution* (pp. 16–37). Oxford University Press.
Pinker, S. (2010). The Cognitive Niche: Coevolution of Intelligence, Sociality, and Language. *PNAS, 107*(11), 8993–8999.

Pisula, W. (forthcoming). Curiosity. In J. Vonk & T. K. Shackelford (Eds.), *Encyclopedia of Animal Cognition and Behavior*. https://doi.org/10.1007/978-3-319-47829-6_2085-1

Putnam, H. (1987). *The Many Faces of Realism*. Open Court.

Putnam, H. (1990). Realism with a Human Face. In H. Putnam (Ed.), *Realism with a Human Face* (edited by J. Conant). Harvard University Press, 3–29.

Rowlands, M. (1999). *The Body in Mind: Understanding Cognitive Processes*. Cambridge University Press.

Rowlands, M., & Monsó, S. (2017). Animals as Reflexive Thinkers: The Aponoian Paradigm. In L. Kalof (Ed.), *The Oxford Handbook of Animal Studies* (pp. 319–341). Oxford University Press.

Ryan, P. A., Powers, S. T., & Watson, R. A. (2016). Social Niche Construction and Evolutionary Transitions on Individuality. *Biology and Philosophy, 31*(1), 59–79.

Shannon, C. E., & Weaver, W. (1949). *The Mathematical Theory of Communication*. University of Illinois Press.

Simons, P. M. (1987). *Parts: A Study in Ontology*. Clarendon Press.

Smith, B. (1996). Mereotopology: A Theory of Parts and Boundaries. *Data & Knowledge Engineering, 20*, 287–303.

Smith, B., & Brogaard, B. (2002). Quantum Mereotopology. *Annals of Mathematics and Artificial Intelligence, 36*, 153–175.

Smith, B., & Varzi, A. (1999). The Niche. *Noûs, 33*(2), 214–238.

Smith, B., & Varzi, A. (2000). Fiat and Bona Fide Boundaries. *Philosophy and Phenomenological Research, 60*(2), 401–420.

Sterelny, K. (2007). Social Intelligence, Human Intelligence and Niche Construction. *Philosophical Transactions of the Royal Society B, 362*, 719–730.

Sterelny, K. (2012). *The Evolved Apprentice: How Evolution Made Humans Unique*. MIT Press.

Stewart, J., Gapenne, O., & Di Paolo, E. A. (Eds.). (2010). *Enaction: Toward a New Paradigm for Cognitive Science*. MIT Press.

Stotz, K. (2010). Human Nature and Cognitive-Developmental Niche Construction. *Phenomenology and the Cognitive Sciences, 9*, 483–501.

Thompson, E. (2007). *Mind in Life: Biology, Phenomenology, and the Sciences of Mind*. Harvard University Press.

Thompson, E., & Stapleteon, M. (2009). Making Sense of Sense-Making: Reflections on Enactive and Extended Mind Theories. *Topoi, 28*(1), 23–30.

Tomasello, M. (2019). *Becoming Human. A Theory of Ontogeny*. Belknap/Harvard University Press.

Tooby, J., & DeVore, I. (1987). The Reconstruction of Hominid Behavioral Evolution Through Strategic Modeling. In W. G. Kinzey (Ed.), *Primate Models of Hominid Behavior* (pp. 183–237). SUNY Press.
van Duijn, M., Keijzer, K., & Franken, D. (2006). Principles of Minimal Cognition: Casting Cognition as Sensorimotor Coordination. *Adaptive Behavior, 14*(2), 157–170.
Varela, F., Thompson, E., & Rosch, E. (1991). *The Embodied Mind: Cognitive Science and Human Experience.* MIT Press.
Varela, F. J. (1995). The Re-enchantment of the Concrete: Some Biological Ingredients for a Nouvelle Cognitive Science. In L. Steels & R. Brooks (Eds.), *The Artificial Life Route to Artificial Intelligence: Building Embodied, Situated Agents* (pp. 11–20). Lawrence Erlbaum.
Varzi, A. (2007). Spatial Reasoning and Ontology: Parts, Wholes, and Locations. In M. Aiello, I. Pratt-Hartmann, & J. van Benthem (Eds.), *Handbook of Spatial Logics* (pp. 945–1038). Springer.
von Foerster, H. (1960). On Self-organizing Systems and Their Environments. In M. C. Yovits & S. Cameron (Eds.), *Self-Organizing Systems* (pp. 31–50). Pergamon Press.
von Uexküll, J. (1926). *Theoretical Biology.* Harcourt, Brace & Co.
Wallach, E. (2016). Niche Construction Theory as an Explanatory Framework for Human Phenomena. *Synthese, 193*(8), 2595–2618.
Washburn, M. (1908). The Animal Mind. A Text-book of Comparative Psychology, New York: Macmillan
Weber, A., & Varela, F. (2002). Life after Kant. Natural Purposes and the Autopoietic Foundations of Biological Individuality. *Phenomenology and the Cognitive Sciences, 1*(2), 97–125.
Werner, K. (2019). Cognitive confinement: theoretical considerations on the construction of a cognitive niche, and on how it can go wrong, Synthese 198 (7): 6297-6328: https://doi.org/10.1007/s11229-019-02464-7
Werner, K. (2020). Enactment and Construction of the Cognitive Niche: Toward an Ontology of the Mind- World Connection. Synthese 197 (3): 1313-1341. https://doi.org/10.1007/s11229-018-1756-1
Zebrowski, R. L. (2010). In Dialogue with the World. Merleau-Ponty, Rodney Brooks, and Embodied Artificial Intelligence. *Journal of Consciousness Studies, 17*(7–8), 156–172.

6

Epilogue: Spandrels and Philosophy

6.1 What Was This Book About? Questions and Answers

Below, I summarize the results of this book in the form of question-answer pairs, which seems especially appropriate given some of the issues that were discussed. I then spill some more ink on the concern that emerged in the last section of Chap. 5.

- *When are beliefs—including those held by philosophers—comparable?*
  Long story short, two beliefs are comparable if and only if they answer the same questions. However, given the fact that questions may also be undetermined to some degree, and that we frequently only learn what actual desideratum is behind the question when we consider possible answers, a fuller account of the puzzle would go as follows: beliefs are ingredients of larger structures involving desiderata (demands for information) and saturata (information that satisfies the demand). Two beliefs are therefore comparable if their embodiments within such larger structures have common elements. To put it as simply as possible: two beliefs are comparable if they answer the same demand for information.

K. Werner, *The Embodied Philosopher*, New Directions in Philosophy and Cognitive Science, https://doi.org/10.1007/978-3-030-79964-9_6

- *How do we manage to come up with questions and question-answers pairs?*
  There are many ways in which one can be led to ask a question: usually, as part of standard scientific practice, one is instructed to ask specific questions that target well-defined gaps in the structure of knowledge. Aside from straightforward instructions, there is also a logical procedure for generating questions consisting in replacing names and predicates with variables, and so on. Yet, if the whole structure of knowledge is supposed to be anchored in our actual surroundings, not just within our representational machinery, there must be a specific cognitive capacity bringing parts of reality into a specific presentation, as the result of which they present themselves as problems. This is the capacity for problematization which throughout this book I have called erotetic intuition. I argued that erotetic intuition is irreducible to instructions or logical procedures, being more primitive than they are and thus more fundamental and mysterious. We lacked an appropriate vocabulary to talk about it in Chap. 3 and therefore covered our ignorance with the intuition talk.

- *What is special about the exercise of erotetic intuition that philosophers act upon?*
  The specificity of the philosophical pursuit of erotetic intuition boils down to what I have termed the recognition of one's epistemic position: awareness of the "fact" that all our "images" of the world—the manifest as well as the scientific—are presentations that fit our cognitive apparatus. I argued that this opens up a level of generality at which we can ask boundary questions. These do not target a specific part or aspect of the world, but the most fundamental features and status of the world-presentation as such. That said, philosophy used to be criticized for transcending the limits of our actual epistemic capacities. There is a legitimate point in this criticism, for the logical status of philosophical claims is, indeed, problematic. Yet the serious problem with this argumentation is that it seems to assume that these alleged limits are somehow fixed, ready-made, and inflexible. How is that a realistic claim given what we know about the evolution of cognition as such, and more specifically about the development of our conceptual frameworks and the history of scientific, philosophical, and other ideas? In Chap. 4 I aimed to make the case that

insofar as there is a limit at every point in history and with respect to every particular setup of cognitive (especially conceptual) tools, there is also a need for an intellectual pursuit dedicated to discovering this limit. For it is not a solid wall, but something more dynamic. Philosophy can be seen as exactly the kind of intellectual endeavor ready to do the job.

- *Is there any way to model the pursuit of problematization in general and philosophical problematization in particular?*

  The key thing about erotetic intuition is that it responds to the need to establish a direct link between the domain of cognition—the space of reasons, to put it in more traditional terms—and the surrounding world. We therefore need a conceptual framework capable of expressing such links more generally, instead of focusing solely on the internal processing of representations. However the intuition talk was just a provisional means to articulate some thoughts. For that reason Chap. 5 directs our attention to the domains of embodied cognition, enactivism, and the biology of cognition; areas that have not been considered in meta-philosophy and the theory of questions. The chapter unpacks in detail a grossly simplified model of the cognitive niche and re-articulates the capacity to problematize. Long story short, problematization can be seen in this context as the capacity to manipulate certain aspects of the cognitive niche, defined as a relatively stable product of niche construction, in order to re-introduce selected entities that otherwise yield no cognitive gains into the active process of niche construction. Seen from this angle, we could metaphorically say that the aim of philosophical problematization is to unveil the structural or "architectural" features of any cognitive niche possible. A more technical rendering of this idea was provided.

6.2 Toward an Anti-reductionist Naturalism About Philosophy

It is a rather uncontroversial claim that the capacity to problematize is beneficial to cognizers, since it enables them to extract information from entities that would otherwise be considered too certain to deliver anything of value, so to speak. There is virtually no entity that could not be transformed into a vehicle of information if a situation makes it beneficial. Now, according to the model proposed, problematization seeks to tap zero-informational objects and constants in the problematized domain in order to elicit what is not constant and can thus potentially convey some information. We can now see that the purpose of a philosophy that prefers to go big and seeks to pin down the fundamental objects and constants is to unveil *an absolute information-zero level of our cognitive niches*. Whether or not it is a doable, such a peculiar kind of pursuit must be a skill like no other. Therefore, going back to my training argument from the final section of Chap. 5, we need to have places, like intellectual gyms, where we are taught to philosophize. Traditional philosophy is the pursuit of radical problematization *in pure form* which must be mastered in order to be properly executed in specific cases.

That said, the capacity to philosophically problematize our cognitive niches is also a product of cognitive evolution—some biological systems acquired this specific ability to develop their problematizing skills in purely abstract form. Now, someone might argue that philosophy is built on a certain crucial adaptation, yet all of its actual content, all these strange theories of everything, are just a byproduct of this useful adaptation and have no value of their own.

This argument reminds me of one of the most widely discussed papers in theoretical biology and philosophy of biology, namely Stephen Jay Gould's and Richard Lewontin's (1979) *The Spandrels of San Marco and the Panglossian Paradigm: A Critique of the Adaptationist Programme*. In this masterpiece of rhetoric, Gould and Lewontin criticize adaptationism understood as a view of evolution according to which a living creature's each and every property must be explained in terms of adaptation and contribution to survival within a specific, pre-given circumstance. They illustrated, or in fact ridiculed, this view by invoking the character Pangloss from Voltaire's "Candide", who preposterously argues that the shape of man's nose was made the way it is to enable man to wear eyeglasses. Hence—Pangloss schooled poor Candide—everything is for a purpose. Yet a more fascinating illustration refers to St. Mark's Basilica in Venice. Gould and Lewontin say that if we were to accept adaptationism in art and architecture, we would have to come to the conclusion that the beautiful mosaic on the spandrels of the basilica was the very reason why these spandrels were erected—their purpose was to make room for the images. However, this is the absurdity we end up with when we take one of the current applications or functions of something and conflate it with the reason why it was brought into existence. For the erection of the spandrels is in fact due to certain internal architectural constraints. If the builders of the basilica wanted to complete a rectangular tower with a dome, they *had to* produce these additional elements which were subsequently used as a support for the mosaics, not the other way around. Analogously, they hold, many structural properties of organisms might have evolved due to certain internal structural constraints, even though some of them would eventually prove useful for other purposes, too. To take an example not mentioned in the paper, Keijzer and Arnellos (2017) and Godfrey-Smith (2016) discuss the claim that the nervous system originally evolved to internally integrate more and more complex, multifunctional bodies, not to deal with environmental complexity—and thus was not an adaptation to the environment in the first place, but a certain "architectural" need which then proved so consequential in evolution.

As we look through this lens at the idea of problematization as a crucial adaptation, and at philosophy as its byproduct, we need to keep the spandrels and the nervous system in mind. First, even if the traditional pursuit of philosophizing is a byproduct, an overuse of an otherwise useful problematization training strategy, you cannot call it a *mere* byproduct. For then you would have to call the mosaics at St. Mark's Basilica a mere byproduct, too. Or, to use an example close to my heart, you would also have to consider the mosaic of St. Francis' Basilica in Krakow, a masterpiece of Secession art covering Gothic walls, as a byproduct. Admittedly, a strange byproduct, for these walls had waited for Stanisław Wyspiański to decorate them from the thirteenth century, when they were erected, until the turn of the nineteenth century, with a devastating fire in between. The time-scale of my example is even more telling, for on the one hand it makes it utterly obvious that some constructions may have different origins than their current standing would suggest, while on the other hand it also shows that we should not be too hasty in issuing a judgment as to whether a given byproduct of evolution or any constructive activity gives rise to something of value. For we may not see this value now, while it might manifest itself in the future or from a different angle. The nervous system may be a telling example.

Therefore, first, whether it is a byproduct or not, traditional philosophizing has value in its own right, just like the mentioned works of art, given its role in the broader culture. Second, if the hypothesis about the origin of the nervous system referenced above is correct, it shows that what was initially a byproduct can in turn become a major factor in evolution or in any other creative process. So, byproduct or not, traditional philosophizing—with all these theories about ideas, forms, beings and Being, transcendentals and transcendental subjects, truths and Truth, and knowledge, justice, and goodness—was and continues to be a major factor in the evolution of the human cognitive (epistemic, scientific, cultural, etc.) niche.

References

Godfrey-Smith, P. (2016). Individuality, Subjectivity, and Minimal Cognition. *Biology and Philosophy, 31*(6), 775–796.

Gould, S. J., & Lewontin, R. (1979). The Spandrels of San Marco and the Panglossian Paradigm: A Critique of the Adaptationist Programme. *Proceedings of the Royal Society of London B, 205*(1161), 581–598.

Keijzer, F., & Arnellos, A. (2017). The Animal Sensorimotor Organization: A Challenge for the Environmental Complexity Thesis. *Biology and Philosophy, 32*(3), 421–441.

Index[1]

[1] Note: Page numbers followed by 'n' refer to notes.

K. Werner, *The Embodied Philosopher*, New Directions in Philosophy and Cognitive Science, https://doi.org/10.1007/978-3-030-79964-9

D

T

The manufacturer's authorised representative in the EU is Springer Nature Customer Service Centre GmbH, Europaplatz 3, 69115 Heidelberg, Germany. If you have any concerns regarding our products, please contact ProductSafety@springernature.com

Printed and bound by CPI Group (UK) Ltd, Croydon, CR0 4YY
30/09/2026
02225698-0006